The Universe,
is it Guiding Our Lives?

✱ ✱ ✱
✱

Fred Bonisch

authorHOUSE®

AuthorHouse™
1663 Liberty Drive, Suite 200
Bloomington, IN 47403
www.authorhouse.com
Phone: 1-800-839-8640

© 2009 Fred Bonisch. All rights reserved.

No part of this book may be reproduced, stored in a retrieval system, or transmitted by any means without the written permission of the author.

First published by AuthorHouse 1/26/2009

ISBN: 978-1-4389-4842-3 (sc)
ISBN: 978-1-4389-4843-0 (hc)

Printed in the United States of America
Bloomington, Indiana

This book is printed on acid-free paper.

Table of Contents

Foreword	vii
Acknowledgments	xiii
Man and the Universe	1
Spirits in God's Service	9
The Possibility of Spiritual Guidance	13
Angels, Spirits, and the Saints	17
Our Perception of Heaven and Hell	26
Our Contact with the Questionable Spirits	29
Unusual Occurrences in Our Lives	32
Recognizing Our Blessings	34
A Miracle of Healing	40
Listening to Our Inner Voice	43
Desires Becoming Reality	46
Coming to America	49
Another Dream Fulfilled	51
A True Act of Kindness	53
The Imbalances of Abundance	56
A Young Boy's Dream	59
Continued Blessings	61
Our Tragedies and Blessings	65
Seeking Intercessions	71
Why Do Things Happen	74
The Feeding of the Hundreds	76
A Timely Intervention	79
Trusting With Reservations	81
Difficulties We Face	85
Could Jesus be Chinese	88

The Good and not so Good Experiences	91
Recalling a Blessed Opportunity	95
A Close Call with Disaster	98
Love as an Important Ingredient	101
Be Careful What You Ask For	104
The Universe of the Unbeliever	106
The Universe as Viewed by the Believers	109
Epilogue	117
References	121

Foreword

Whether the universe is possibly guiding or assisting our lives, seems like the ultimate question. It is not clear to me why I feel so inspired to write about this topic, especially considering that I am not a member of the clergy, a scientist, or an astronomer. Although I have a strong curiosity about this issue, I claim no special spiritual insights that would qualify me as an expert on this nearly unanswerable topic. I consider myself an average human being at best. By not claiming any special credentials, I can address this subject from the perspective of a novice and with an open and curious mind. Certainly, whatever I have to say will likely be criticized by the scientific community, the church fathers or both. Being criticized, however, is not what I fear, as that is something every writer must face. Even asking a question of this nature openly, only a few hundred years ago, would have branded the curious one as a heretic. Worse, the offender would have found himself burned at the stake as a result of such liberties. With the realization that our conscience is the ultimate judge of our faith, we are no longer bound by the religion of our birth but are free to choose. This conscious guided freedom also enables us to search for answers previously blocked by strictly enforced church rules. We still argue and debate about whom of us belong to the only true faith group. At least now we have the freedom to ask and search. Surely some may find the title of this book objectionable, but who then is the authority that can possibly give us the answer or dispute our assumptions? Even from among the scientists or the religious, we may be hard pressed to get a concrete response. With this realization, we may have to ultimately concede and accept that we are possibly dealing with something which we are not meant to understand.

It seems that in the end we each have to draw our own conclusions, realizing that perhaps nobody can answer this dynamic question for certain or completely. Surely some of the religious leaders may attempt to provide answers and they too will vary, as each of the world's major religions has a different perception of what lies beyond our understanding. Although the majority of them claim to believe in the same God, each is certain they are the only true believer. This diversity exemplifies different ways in which we view and relate to God. These groups may agree with the general concept that God is the creator and so the center of the universe. A consensus, however, as to whether the universe is possibly guiding our lives is rather unlikely. Certainly people of every generation have asked this question or at least thought about it. It is only more recently that we have come to feel free enough to have an opinion, even if it's questionable or turns out to be wrong.

We need only to think of the faith of the Astronomer and Physicist Galileo Galilei for supporting Nicolaus Copernicus, founder of the Heliocentric Planetary theory. He dared to state that the planets, including Earth, revolve around the sun. We know for certain today that this is correct. This, however, was contrary to the general belief that the Earth was the center of the universe and that all else revolved around our planet. Galileo was subsequently convicted of heresy and sentenced to life in prison by the Inquisition in 1663. In view of his advanced age, his sentence was later modified to house arrest. It clearly demonstrates the power held by the church over many centuries and its eagerness to inflict punishment even in such non-religious matters as astronomy. Realizing how far we have come in such matters, I have attempted to pursue this great question and am hopeful that the reader, too, will accept this with an open mind.

The attempt of this book is by no means intended as evidence or proof, rather its purpose is to simply consider the possibility of the universe having life that can engage with us. We are acquainted with such terms as Heaven and Hell, but how serious are we about these unknown and invisible places? Is the universe just an empty space aside from the occasional planet? What makes our planet so special and why are we here? Do we have a purpose or are we just drifting through a period of time? These are all important questions that have probably been asked individually and from different perspectives. Yet when we consider the reasons for the inequalities that exist around the world, we begin to stumble and even the most renowned religious leaders are usually at a loss for reasonable answers.

To avoid giving the wrong impressions, I do not claim to be a prophet. I have no special insights nor have I ever had contact with any supernatural beings. I have not been blessed with special abilities nor do I feel that my relationship to God, although personal, is not so different from that of everyone else. By having arrived at these conclusions, I wish to approach this subject from the perspective of a humble human being, hoping at least to come to a better understanding as a result of my curiosity. What mostly inspired me to even think about a subject of this magnitude were the numerous and unusual events that have occurred in my own life. When I consider my beginning and where life has taken me, I have come to realize that many steps along my journey were often extraordinary if not miraculous. I realize that this is quite a claim to make and yet I believe that we all have unexplainable events that have occurred in our lives. Haven't we all occasionally exclaimed, "Whoa, where did this come from?" when something unexpected occurred at just the appropriate moment?

I was born and raised in Germany and so my journey may well have had more twists and obstacles then most. By coming to the USA as a young man of twenty, I had to adapt to a new language and a somewhat different culture. I consider these valuable lessons as they have added much to the overall experiences of my life. It is also the recollections of this unusual journey that gave me the added inspiration to write about them.

In my previous book titled *Children of Our Own War, A Boy's Journey* I recalled my family's experiences as we lived through the fearful bombings of World War II in northern Germany. Although this second book is not intended as an autobiography, it describes segments and special events of my life since so many of them have caused me to become inspired and to write about them. I have wondered numerous times whether unseen spirit beings are there to provide assistance and help me. While such thoughts could be fearful, I have usually regarded them with a certain fascination but also with a degree of reverence and respect.

Once, while I was confronted with a personal issue, I came upon a note that read: "Trust your intuition. The universe is guiding your life." This little note impacted me profoundly in resolving the issue before me. It has remained with me since and has become the inspiration for the title of this book.

My religion and my faith have also contributed to my desire to explore this unique and mysterious topic. I wish to caution the reader again that this book is not intended as proof or concrete evidence. Rather it is meant as a beginning, whereby we can simply wonder about whether the universe has a purpose to benefit us. If so, what would be its motivation and how do the benefits manifest themselves? How do we view God's intervention and his possible spirit world in all of this? I am aware that the questions can not be fully answered but by realizing the possibilities, I feel challenged to at least review and to explore them in some detail.

The reader will be able to determine that the general approach I have taken in my writing is that of a Christian. This is by no means intended to offend any other faith group or non-believers. It is simply a direction in which I feel this topic can possibly be pursued and hopefully, to some degree, is found worthy of consideration. It needs to be acknowledged that using this approach, however, is not without its conflicts as it is not acceptable to all. Those of the Protestant faith, including Bible churches, will usually only accept what is scripturally verifiable. Catholics, although believing in scripture as well, also accept the rulings of their hierarchy. These rulings are commonly known as Cannon Law, past practices, and the decisions of its magisterium or councils in the belief that they are inspired by the Holy Spirit. Since these rulings may not necessarily be based on scripture, they are generally known as church laws. Catholics have accepted such rulings over the centuries. This difference, however, has substantially influenced how its followers practice their faith and so it sets them apart from other Christian groups. There are many others of the Christian faith who may not even accept what is considered by most as basic scriptural theology. This reality further complicates any attempt of finding a common ground on the role of the universe from a Christian perspective.

My religious background is that of a Catholic. Although I have tried not to be influenced by my affiliation, I have found that it offers an avenue in which my search seems most practical. Any biblical references which I have made, however, have been taken from the New King James Version which is most commonly used by non-Catholics. By having used this somewhat neutral approach, I am hopeful that it will find a greater acceptance within the Christian community and beyond. The occasional use of Bible references are intended only to support my efforts rather then to claim myself as a biblical scholar.

Much of what is covered in this book is based on true personal experiences. The circumstances under which they occurred have often left me wondering about who it is out there that is guiding me or any of us. These personal experiences set us apart as unique individuals with special purposes and directions.

Acknowledgments

Before I decided to write about this somewhat controversial and mysterious topic, I sought the views and the opinions of both religious and non-religious individuals by means of informal discussions. While their responses about how they viewed the influence of the universe varied, it indicated to me that most of them had already contemplated this issue in their own minds. While they will remain nameless, I am grateful for their comments and opinions.

I wish to express a special thanks to my friends Joanne, Norm, and Barbara for their valuable support with the editing, constructive suggestions, and honest critique. The time that each of them has given this project, is a sign of real friendship which is something I value greatly.

I also wish to thank my children Kimberly, Marc, and Megan for their continued support and encouragement which keeps me writing.

Man and the Universe

The question, whether the universe is guiding our lives, has been nagging me for most of my adult life. I am certain that many others have either thought about or asked this question in one form or another. As human beings we have a curiosity to know what lies beyond our small planet, Earth. We have set foot on the Moon, have landed a space craft on Mars and we continue to venture farther and farther into space. Voyager I space probe, for example, was lunched in 1977 and its mission was to explore the solar system. It has since passed Jupiter and Saturn and is now the most distant man-made object from Earth. Still going strong, Voyager I is now well beyond our own solar system. It continues its journey and still sends daily information to us on Earth. Due to its current distance of about 8.7 billion miles from the sun, it takes about nine hours to receive these daily transmissions. While we have to acknowledge the enormous achievements that have been made by our own scientists, thus far, we have not encountered other living beings on these explorations or at least none we have been told about. The scientific community is optimistic however, that perhaps some day we will encounter planets that can possibly sustain life as we know it.

Numerous nations around the world are in a race trying to explore the unknown of the universe. Much of this may be driven by national pride and the quest to be first. We really need to ask ourselves, what is it that we are hoping to find? Although we are uncertain whether we will ever encounter other living beings, we are, however, fascinated by the fact that the universe has an order that is extremely precise in its movements and timing. By realizing this important factor, how can we not ask ourselves, whether it was created to serve a greater purpose? If so, who then was its creator?

The idea of looking to the universe for answers has been in practice for thousands of years. Often emperors or kings relied on their astronomers or astrologers to check the stars or the position of the planets to help them make major decisions. Whether to engage in a battle on a particular day was often left to these star gazing wizards. Even the three wise men who came from afar to visit the baby Jesus in Bethlehem claimed to have seen the message of a newborn king in the star they followed. We could say that all of this is part of the superstitious past and yet most major newspapers will have a daily horoscope section of predictions based on readings of the stars. Obviously many of its readers still are interested and influenced by these predictions. We can see that our curiosity about the universe has hardly diminished.

Until it's proven otherwise, we are the only known life-sustaining planet in the universe. Some in the past have even believed that our Planet Earth is the center of the universe; this notion, however, has long been proven to be wrong. Regardless of this somewhat disappointing finding, the fact that our planet can and does sustain life should rank us at least as an important element within what seems to us to be an endless universe.

Although we may consider our planet to be important we are still only a microcosm in the vastness of the solar system. Scientists recently announced the finding of a black hole that exists between two galaxies 1.4 billion light years from Earth. It is significant that all the planets, including our own, move about each other in perfect harmony and with a precision that is simply astonishing. We may wonder whether all of this could have possibly come about by itself. Scientists frequently refer to the Big Bang Theory to explain the coming into being of our planet, Earth, some 4.5 billion years ago. The origin of man, according to DNA evidence, is believed to have occurred only as recently as two hundred thousand years ago. Others, however, believe that it was much more recently, approximately a hundred thousand years ago. Some Christians even believe that the world is only 5,000 to 6,000 years old. Here we often find varying opinions between that of the scientific world and those who hold to the Biblical Creation story as described in the book of Genesis.

It is interesting to note that there are several theories that attempt to explain the origins of our planet Earth and/or the universe as a whole. We are more concerned about the question of how humans, the other animals, and plant live came into existence. Initially, Darwin's theory of evolution greatly upset much of Christianity regarding the belief in God as the creator of a complete human being along with the animals and plants. While we view evolution as the scientific approach to deal with this issue, most Christians and Jews no longer find it contradictory to their beliefs. Other orthodox religious hardliners hold firm to the belief of creationism by believing that all that exists suddenly appeared as we see it today. They are clear in their belief that a Creator, namely God, created the universe, with all that it contains, over a 6-day span.

Eugenie C. Scott, Ph.D., Executive Director, National Center for Science Education, Inc. in her paper titled 'But I Don't Believe in Evolution' points out that "Thus except for creationism in the narrow sense, there is no hard and fast line between 'creationists' and "evolutionists'. The theological possibility of God's creating through evolution means the important question in the creation/evolution controversy is not *whether* God created, but *how*."

While the concepts of evolution and creationism may be viewed differently, it should be noted that by accepting evolution, science and faith are not opposing each other. Creationism is a faith-based assumption in which science attempts to play no part. In fact the concept of creationism is accepted by Christians, Jews, and Muslims alike.

Most major religions believe in the existence of a higher power, a Creator or God, who is above all. Contrary to the rejection of this notion by atheists, the advantage of holding to the belief of a creator, a God, is that one does not really need to understand all the mysteries of the universe. One simply needs to trust and believe that mankind is in good and responsible hands.

If we believe and accept that God is the creator of the universe, then we too must be part of His creation. It would, however, seem unreasonable to think that the entire universe was created just for mankind alone. If we assume this to be the case, we may want to ask, who else is out there and why have we not been able to make contact with them?

Many well-respected people claim to have had contact with beings from outer space this includes sightings of so-called flying saucers. Some even claim to have been captured by alien beings and describe their experiences in detail. The fact that the scientific community is at odds with the existence and the validity of these sightings leaves this issue simply unanswered and therefore somewhat doubtful. The church has generally been careful not to take an official position in regard to these occurrences. A number of Christians have written about this and some have even suggested that UFO's are a product of the demonic. This, however, provides no significant input that would alter the general assumptions and position of the scientific community.

While recognizing the reality that we are in fact only a microcosm in the universe, it would seem wrong to think that the rest is simply wasted space. For example, instead of searching for visible life outside our small planet, should we consider the possibility that the universe as giving life support to us on earth? We already know that the sun provides us with light and life-sustaining warmth. Wind and rain needed for plant growth move above the earth's surface aiding our survival. As the Earth rotates with perfect timing, it provides us with night and day giving us a daily period of rest and a time for our daily activities. Contrary to other planets encountered so far, the Earth provides life and the means to sustain it. More could be said about this but it is sufficient to realize that our planet is truly unique in so many respects.

To assume that the rest of the universe is just an empty space for no other purpose than to contain the occasional planet or galaxy seems shortsighted. As we look at our own planet, there is already a great deal of concern about the fact that perhaps someday it may become overpopulated and no longer be able to feed all its inhabitants. Starvation is already a reality in many parts of the world with little hope for improvement. Pollution and the fear of the effects of global warming have become serious byproducts of our growing society. Instant communication has made us aware of enormous unrest and the ability to destroy all forms of life on Earth within just hours.

Faced with this reality, we may wonder whether other life sustaining planets have become extinct through similar occurrences of self destruction. When we begin to ask whether we are alone in this predicament, we won't receive a clear response from anyone. This, however, should not deter us from at least considering the possibility that we are sharing the universe with other beings.

If we accept that we are a creation of God, why then would it seem so unthinkable that the same God could create other beings and place them into the same universe with us? We can only speculate about such a possibility, but should there be other beings out there, they too could serve a purpose known only to the Creator. With our limited human capacity to perceive, we can only recognize solid bodies and objects and so remain skeptical of anything invisible. Since we have not been able to encounter other beings with solid bodies, similar to ourselves, perhaps we ought to consider the possibility of life forms invisible to us. If, in fact, such a spirit population exists, it too will have limitations which prevent us from openly communicating with each other.

By exploring such a possibility further, we need to be cognizant of the possibility that we are physical beings living alongside spiritual beings. We cannot see them but we can try communicating to them. They, on the other hand, may likely be able to see and hear us but lack the ability to verbally communicate with us, or so we may wish to believe.

By allowing ourselves to consider the above concept of our sharing the universe with such spiritual beings, it may be of interest to explore possible commonalities and purposes. We have already acknowledged that we are both created by God and so share a common origin. Where we begin to differ is in the fact that our earthly boundaries are very well defined by solids, liquids and gases. Our bodies are conditioned to live on Earth. We find it impossible to exist even beyond just ten thousand feet into space without special equipment to provide us with needed pressure and oxygen. Gravity, like a magnetic pull, keeps us earthbound and prevents us from drifting out into space. As human beings or earthlings, we are born and are destined to die. We need air, food, and liquids to sustain us. But equally, we need the association of other human beings and have a desire to love and to be loved.

Fred Bonisch

Our existence depends greatly on the interaction with other human beings.

In spite of our physical nature, we too have a spirituality. We call it our soul. It is something which no one has been able to describe or to define. It is that unseen portion that becomes part of our being at conception and returns to the Creator when our life on Earth is completed. Our bodies, necessary for our earthly existence, become the dwelling places of our souls.

Although we are mostly unaware of its existence or where it is located within our bodies, it is there nonetheless and perhaps it is our connection to God himself. Other than just by speculations, we are unable to define how it interacts in our daily lives. Since it is a spiritual entity, it may possibly, but not necessarily, guide our feelings and emotions as these are distinctly different from the physical makeup of our bodies. By our natural ability to determine much of what is right and wrong, our soul may well be a guiding element as well. It is interesting to note that the soul and its vulnerability are referred to in the Bible over thirty times. Just a few examples are: Deuteronomy 6:5, Proverbs 6:32, Proverbs 8:36, and 2 Peter 2:8. Perhaps the most specific of these is shown in Matthew 10:28 from Jesus himself when he cautions the believers, "And do not fear those who kill the body but cannot kill the soul. But rather fear Him who is able to destroy both soul and body in hell." (The New King James Version) From this warning we can see that the body and soul are separate entities and yet they can also be lost or destroyed together. The wording of different Bibles may vary, but the concept is held as valid by at least the majority of the Christian world. This, however, is not shared equally by other religions and should, therefore, not be viewed as a universal concept.

We may also wonder whether our soul remains constant from the time of conception to our moment of death or whether it matures as we mature as human beings. Again, we can only speculate about this secret spirit within us, as even the greatest theological minds have pondered about this mystery over the centuries. Perhaps it is the breath of God that comes to us at the time of conception and returns to Him as our last breath when we leave this Earth.

What makes our soul so unrecognizable is the fact that it is invisible. With our inability to fully understand its function, we assume that it, too, is limited in ability to communicate with the universe around us. Although this is purely speculation and no proof exists, perhaps our soul resides in our sub-conscious mind which becomes activated during our sleeping and sub-conscious hours. Our ability to dream so vividly, which occasionally allows us to have visual and verbal contact with our dearly departed, makes this a possibility at least worth considering. This then might cause us to question our ability to communicate with the spirit world on a sub-conscious level.

It should be noted, however, that according to Deuteronomy 18:11, Christians are forbidden from consciously attempting to contact the dead. It also warns the believers about consulting a medium or a spiritualist or those who call upon the dead. By cautioning the believers in this manner, the Bible, in fact, acknowledges the existence of a spirit world while at the same time warning them of the influence that evil spirits could exert on us. The evil spirits referred to here are assumed to be the fallen angels who have joined Lucifer, also known as Satan or Devil, Lucifer once held a prominent position among God's angels. Lucifer and his fallen angels represent the dark side of the spirit world and are viewed as having values contrary to those we hold as followers of God. It is interesting to note that while the Catholic Church teaches this doctrine as well, it also encourages its followers to call upon the saints for the purpose of interceding with God on their behalf. One might question whether this is in fact a contradiction between its teaching and its practice. Certainly there are different viewpoints even among the world's Christians. For those who call upon the saints, it has often been a great source of comfort and hope. As this practice also leads us into the positive aspect of a possible spirit world, more discussion will follow in a later chapter.

Being aware of these varying views should not deter us from assuming that our soul becomes that spiritual commonality we share with the spirit world around us, provided of course, we believe that such a world exists. By accepting this fact, we need to recognize that we have the freedom to decide which side of the spirit world we choose to be on.

Surely, all of us are familiar with the concept of intuition. It is an ability to sense something that has not yet occurred visibly. It is a concept recognized both in psychology and philosophy. Prophets are often intuitive when they warn their audiences of what will be. Here, too, is something we occasionally experience whether it may be nearby or far away. Twins are known to have felt each other's feelings regardless of the distance between them. We sometimes find that people in love, but separated from each other, have sensed a special event one of them experienced. Some people claim a great gift of intuition; others take little note or ignore this ability altogether. It should cause us to wonder about who or what allows this invisible transmission to occur. Why is it just sometimes and not continuous? Does it engage a spirituality that we possess but are unaware of? Does our soul interact in such a situation? Some express the belief that mankind once had the ability to transmit thoughts over any distance but we have lost this great gift over time. Certainly that is just conjecture without any basis but an interesting thought nonetheless. It may, however, cause us to wonder about whether we have a connection to the universe that enables us to experience intuition regardless of how minute it may seem.

Spirits in God's Service

If, on the other hand, we dare to attempt to describe the spiritual, it may serve us to, at least, consider Webster's definition of spirits. He simply describes them as: "Supernatural Beings," a rather broad term which, other than stirring our imagination, gives us little to go on. For most of us, the term supernatural automatically implies powers or abilities well beyond our own. If we try to imagine the existence of a purely spiritual world, we can, of course, only imagine and speculate what it may possibly be like. To begin with we would need to make some very basic assumptions. Spirits, as we think of them, would have no need for the things that sustain humans. Once established, their lifespan could be indefinite or as God desires. They are not limited by space, time, or earthly bodies and so can travel the universe in an instant. These abilities are far too difficult for our human minds to imagine. Just thoughts alone become their means of communication; something well beyond our own limitations.

In spite of the fact that we recognize our differences, we keep coming back to the fact that we both originated through the will of God. This becomes an important factor when we dare to speculate about His intentions. In Scripture (John 3:16, The New King James Version) we are told that: *"For God so loved the world that He gave His only begotten Son, that whoever believes in Him should not perish but have everlasting life."* Christians view this as a sign of God's abundant love and concern for mankind. As the Creator of all that exists, we consider ourselves His children. As such, we hope that He will not leave us alone and defenseless. This also assures Believers of an everlasting life in the presence of God. This is an important factor and promise as we consider the possibility of a spirit world that is assured to live for all eternity. While they enjoy living in the presence of God, it would seem possible that these spirits would also be able to assist mankind in some manner. It may not even be so unreasonable to think that watching over us by means of this presumed spirit world could well be a continued gift of His love for us.

Much of the Christian world believes in the existence of the Holy Spirit as part of God's holy Trinity and so the concept of spirits leads us directly to God who is a spirit Himself. It is, in fact, the understanding of believers that the Holy Spirit leads the church and that he is our source of guidance and understanding. By accepting that God is a spirit being, although he can take on any form he chooses, we can begin to understand that others who live in His presence need to be spirit beings as well. I wish to caution the reader that this is by no means intended as proof of the existence of a spirit world. Rather, it points out that the concept of relying on God for spiritual guidance is part of the very foundation of our Christianity and other religions.

We need to be careful not to confuse guiding with controlling. We know that God gave us a free will, but He has also invited us to call on Him for all of our needs. Although we lack the ability to communicate with this so called spirit world, we trust nonetheless that our prayers are heard. Since we usually receive no oral responses, we can easily think of this as a one-way communication. We are assured by Jesus himself (John 15:16) that our prayers reach our Heavenly Father. Although this is not based on biblical evidence, our prayers may well be heard by other spirit beings as well. If we assume this to be so, then perhaps it is here where we may need to acknowledge a superior ability by the spirits surrounding us in that they are likely able to hear and monitor our movements and needs.

This is by no means to suggest that God has a need of such spiritual helpers. Rather, He may have chosen them for their own continued spiritual growth in His service. I could well imagine that even while they enjoy being in the presence of God, the spirits may still aim to perfect their existence in their faithful service to Him.

The Universe, is it Guiding Our Lives?

As believers we need to be careful, however, not to assume that their service makes us less important especially when we take into account that we are created in the image of God. I assume this to mean that we have a certain likeness but are definitely not equal. If, in fact, we are that important to Him, it becomes inconceivable to think that we are placed on this Earth without a higher plan or purpose. As human beings we lack the ability to foresee our individual future and so are unaware of His special plan for us. Books have been written about this very question by various people. " *The Search for Meaning*" by Dennis Ford and "THE LATEST ANSWERS to the OLDEST QUESTIONS " by Nicholas Fearn are books that have approached this issue from a philosophical perspective. In his book titled, "A NEW EARTH *Awakening to Your Life's Purpose*" spiritual teacher Eckhart Tolle concerns himself with the state and the transformation of consciousness. Rick Warren, in his New York Times bestseller "THE PURPOSE DRIVEN Life", attempts to answer the question "What on earth am I here for?" His approach is that of a Christian and so defines the purpose for our existence from that perspective. No matter how we attempt to define our purpose as it relates to God, neighbor, and family, each of us has a roll to fulfill which is ultimately known only to God Himself.

Even some of the most respected people in history have asked themselves "What am I here for?" I have come across many older people who are still trying to find an answer to this very basic question and so wonder whether they have fulfilled or are fulfilling their purpose. Others have placed themselves on a self-imposed hold, waiting for their real calling in the belief that they are intended for a higher purpose. If, on the other hand, we assume that we may have already fulfilled our purpose or purposes, this would make the rest of our lives nearly meaningless and empty. Even in our old age, our purpose may simply be to exist for others who still need and love us. We usually tend to think of our purpose in the singular; when, in fact, we probably all have multiple roles - each serving different purposes and functions.

Recently, while I listened to a sermon, the speaker referred to us as having specific roles to fulfill. He pointed out that each of our individual roles has a purpose which affects others or even the world at large in some way. In trying to emphasize his point, he asked the group to consider where they thought the civil rights movement would be today without the voice and actions of the former civil rights leader Dr. Martin Luther King Jr. Obviously, the answers were self evident. We may not see our role as that significant. Without the ability to look into the future, we are unable to predict the impact of our actions. If, however, we fail to fulfill our designated rolls, we are likely to leave a void that can have negative implications for others.

This may be an appropriate time to examine two important issues; namely, God's plan and the spirit world in it. I firmly believe that God determines the plan or purpose for each of our lives and I take great comfort in that fact. Shortly after writing this paragraph, I happened to drive by my own church where at its entrance a large sign read: "*Trust in God's Plan.*" Reading this sign was like a message of reassurance and timely confirmation of what I am trying to convey. If we approach this from the position of believers, we need to acknowledge that God is in control first and foremost. The spirit world then becomes God's instrument for whatever purpose he intends. While realizing that we have no official proof for this, I like to believe and trust that the good spirits carry out his work by helping us with our daily lives. Being open and trusting in this invisible spiritual force should be a comfort; we know that we are never alone.

If we trust that we are supported by this unseen force, it would seem that the spirits may have knowledge of the will of God regarding our individual circumstances. Further, they have been given certain powers or abilities to act on His behalf and for our benefit. I need to again caution the reader that there is no official proof of these assumptions. Coming from a Catholic background, engaging the saints through prayer and petition has always been a standard practice. This, however, is not an accepted practice by many Christian or other religious groups.

Atheists generally deny the existence of God, yet, aside from our Christian belief, the simple words *In God We Trust* were added to our money in 1864 and have been the guiding words of our nation ever since. When we begin to reject our God and creator, the universe becomes an empty and meaningless space. We, too, become just drifters, condemned to a meaningless life without hope. For us believers, regardless of our religion, *In God We Trust*, gives meaning to life and a hope for salvation.

The Possibility of Spiritual Guidance

When we take time to look back on our lives, we find that we have each experienced strange and unexplainable occurrences that astonished us but were usually termed intuition, coincidence, luck, being at the right place at the right time, etc. Usually in such circumstances we desire to take credit for how lucky or how smart we are. I have had many such situations throughout my life and have come to accept them as comforting, knowing that someone is watching over me and guiding me. Not all of the events I have experienced were the result of my asking for them. They have usually occurred simply because I had a specific need. I find it astonishing, that a spirit being, unknown to me, can be so in tune with my life that it can monitor my needs and hopes and so can act to my benefit.

I like to think that the contact we have with our spiritual guide occurs more frequently than most of us ever realize. They are the ones who choose the means by which to communicate or to assist us, often by using the least obvious route. Every once in a while we experience a special moment that causes us to think and take note. We may think of a particular person and suddenly the phone rings and that person is calling us. Other times we wonder why suddenly a person from long ago is on our mind. Does the universe become the connecting link in such cases by simply passing on thoughts of that special someone?

Fred Bonisch

Recently during a television interview the respected actor and author Sidney Portier, expressed his strong belief in what he called the forces of the universe. He even went as far as to credit these forces with intervening in his early childhood. He recalled things from his childhood that were so borderline negative that it could have easily pulled him into a path of disaster rather then onto the road of success and respect. He described this tilting point as so narrow, that without the help of these unseen forces he might not have made it on his own. While I agree with him about the energy that is out there, I rather credit this force as coming from God, the Creator, and those who serve Him.

As a result of my interest in this topic, I have come to know people who firmly believe that the spirits often communicate with us in our dreams. I recall that shortly after the passing of my father, he appeared to Mom in a dream. In it he told her that it was not her time to go yet and that she would be here on Earth for a long time. He also encouraged her not to worry as he would continue to take care of her. That was nearly thirty years ago and Mom is now ninety-six. To the skeptics I wish to say that both of these promises are still being fulfilled. I am certain that many of the readers, including religious groups, will reject this assumption as nonsense or even demonic. I trust or rather am certain that my father, too, has become part of that saintly spirit world and that it was his unique way of communicating with Mom. Her telling us of this timely dream, was like a message of assurance that he is still in touch with us. I trust that with this new ability to experience the other side, he continues to be there for his earthly family. With his new perspective of our earthly existence and our spiritual future, his interaction with our lives may not always be as we would expect and so it will often go unrecognized.

Hopefully, we have all experienced love in one form or another. For most of us it came in its most natural form within the family unit. It begins at birth and remains with us throughout our entire lifetime. It does not even stop at death since we who are left behind continue to love our dearly departed. It is not a feeling that we can simply turn off just because someone is no longer with us in the physical sense. I feel confident that although the departed loved one has become part of the spiritual world, their bond of love continues for us as well.

The Universe, is it Guiding Our Lives?

From Mitch Albom's book, *Five People You Meet In Heaven*, we get a somewhat different perspective of the spirits we may encounter after our own passing. He talks about meeting the spirits of people who in some way affected or influenced our earthly life but whose encounters we may not necessarily remember. We need to keep in mind that his book is purely fictional and so will not be accepted as credible evidence. The number of copies of his book that have been sold indicates our strong desire to speculate about this unknown area. While he talks about those who may possibly meet us after our own death, I would like to explore the possibility of a relationship with those spirits while we are still here on earth. Although his and my own view of the spirits and their roles towards us may differ, we both acknowledge their possible existence.

I must admit that in spite of my belief in the existence of such a spirit world, I can not claim success in even one single personal experience as I attempted to make contact with them. I can't imagine what my reaction would be should I ever see or even hear a spirit's reply. Although we don't expect a reply, there is still that twinkle of hope that maybe just once we will receive some kind of response. Let's face it, who would not like to feel that brush of air or the movement of an object as their gentle acknowledgment of our attempts. Perhaps that is their unique ability not only to be unseen but to respond in the most inconspicuous manner, where we on the other hand, need physical proof and preferably with a time-table attached.

It seems unrealistic, however, to think that we can just stumble from one coincidence to the next by simply by sheer luck or through our own ability. If we truly claim to be part of the universe, it should not be so surprising that the universe may possibly be there to help us. Considering such a possibility, it would appear that we have little, other than perhaps pollution, to give in return. The theory of the universe as our guide and support can not be proven in a physical sense. Although we pointed out that the sun provides life support, we recognize it as something constant, physical, and unchanging as it carries out its functions within the universe.

What I hope to do is to concentrate on the unseen, the spiritual portion of the universe. To consider its existence, we need to draw on our experiences. Through our prayers and thoughts we may well already be in touch with it more frequently than we realize. I am hopeful that by sharing some of my own experiences, I can at least inspire my readers to become more in tune with the special occurrences in their own lives. By doing so, maybe some will begin to consider the possibility of such an outside influence. It should make us realize the comfort it can provide; knowing that we have trusted friends from within the universe who are always there to help and assist us.

Certainly many will reject this notion as simply nonsense; such a reaction is understandable since this concept is somewhat difficult to prove. We need to realize, however, that there are other things in which we believe without having ever seen them.

Many of us believe in the existence of God and a life hereafter yet none of us has ever seen or had the experience. The fact that we continue to breathe without thinking about it is a mystery we accept without giving much thought to it. It is in our nature to accept what we feel suits us and reject what we don't understand. Certainly many will reject the notion of a spirit world, while others who consider its possibility may begin to recognize its influence in their own lives. Whether we accept or reject this notion may not greatly alter our lives, but trusting that we are not walking alone should be a comfort especially to those who live lonely or solitary lives.

We can hopefully begin to see that by addressing this issue from a religious or believer's perspective, it provides us with an avenue not available from the scientific community. Instead of insisting on concrete or proven evidence, we approach it from the position of faith. Even here we need to be conscious of the sensitivity of this issue and how it may affect those of various religious and non-religious backgrounds. It should be remembered that we are trying to explore an issue that was a topic that was of taboo for more then fifteen hundred years. Even now we need to view this from the perspective of an opportunity that may serve us rather than as something true and factual.

Angels, Spirits, and the Saints

The purpose here is not to pay a great deal of attention to what we usually refer to as bad spirits. Unfortunately, too often they receive more publicity because we usually associate them with mysteries and special powers. On the other hand we should not ignore their existence as they often pull us away from what is good. It's the subtle urges that we all feel to do things against our conscience or better judgment. Here is where our good spirits can be of the greatest value by keeping us on the course for which we have been chosen. This then could be a function they perform; assisting us to counteract the forces of evil every one of us faces on an almost daily basis. We need to recall that while they may assist us, the choices we make continue to be our own.

We are all familiar with such entities as poltergeists or restless ghosts that usually haunt old houses. They have been described as spirits who have not yet found peace and so are still trapped between this life and the final spirit world of our understanding. Although we usually dismiss this as rather doubtful, again the Catholic Church has specially designated priests who are trained to perform exorcisms in order to dispel questionable spirits and demons. The Orthodox Christians view spirits/demons that are exorcized as fallen angels that have always been angels or demons. Since this exorcism practice is now seldom performed, it is usually done without publicity or fanfare and so is rarely mentioned. The fact, however, that this practice is still performed by the church gives this restless spirit or demon belief a certain credibility.

Many will say yes, we know about the spirit world and we call them angels. It is important to recall that God created the angels for a specific purpose. Angels were not born as humans and then later became angels or at least so we believe. We have been taught that the angels were created for the purpose of serving God. From the Bible we see that angels often became the messengers of God. We see this in the story of Mary (Luke 1:26-38), for example. From the story of the three angels visiting Abraham and telling him that he and Sarah would have a son in their old age, we can see their ability to take on human form. We have also been taught that there are fallen angels. The most notorious one is known as Lucifer. We look upon him as the Devil whose quest it is to conquer our souls. From this it would seem that angels, although created by God, had at least at one time been given the ability to choose between good and evil, (not so different from ourselves).

While I recently had an opportunity to discus this topic with a clergy friend, he explained that once angels have fallen from grace, there is no return. Man on the other hand will always be welcomed back by God as a repentant sinner. If we assume this to be so, we may ask whether the spirits or the saints are also subject to falling from grace. Knowing, however, that they live in the presence and glory of God and are able to see those condemned to what we think of as hell for eternity, seems to be a sufficient deterrent to straying away from the love and glory of God.

Having learned about such terms as archangels, we must assume that there exists a hierarchy among the angels as well. In the Gospel of Luke (Luke 1:11-38) the angel Gabriel brings a message from God to both Zacharias and Mary, the mother of Jesus. In (Luke 1:19) he declares that: "…..*I am Gabriel, who stands in the presence of God, and was sent to speak to you and bring you these glad tidings*….." We usually see angels and archangels pictured in varying forms. They can appear as little boys with wings, stern male figures with larger wings, or even holding swords indicating their role as protectors or executioners. Angels performing different roles in the service of God are mentioned throughout the Old Testament (Numbers 22:23, 22:33, 1 Chronicles

The Universe, is it Guiding Our Lives?

21:12, 21:27, 30, Isaiah 6, etc.) Although we really don't know enough about angels to give specific details, we can assume that they are, in fact, created to serve God and can adapt to whatever purpose He chooses.

Most of us have been taught that we each have a guardian angel who walks and protects us from harm. If we truly hold to this belief, how then do we justify this theory when we see so many innocent children around the world being hurt, abused or even killed? This is not intended to judge the work of the angels but rather to question what their real purpose may be. Surely this will be received with mixed feelings but I trust that the purpose of the angels is reserved to serving God alone. Scripture has presented them in numerous ways while carrying out God's will. At times we have seen them described as messengers or protectors but we have also heard of them slaying every first born Egyptian during the time of the Passover. Here we see a possible distinction in that the angels can carry out God's work without conscience. Although we don't know this for certain, spirit beings of human origin have a conscience as part of their soul. It is something we have here on Earth and I trust it will follow us into the next life. It is that ability that has created us in the image of God, something Adam and Eve had to learn the hard way. We have never heard or read that angels were created in the image of God but rather only to serve him. It would seem an important distinction when we consider which of the two categories would be better suited to give us at least occasional guidance. By our identical origins, the spirits may well have an advantage in assisting us.

Spirit beings are certainly a separate category. I like to think of them as departed humans who through dying have transitioned into what we often refer to as eternity. They now exist in the presence of God and serve and glorify Him without the burden of their earthly bodies and the problems associated with them. It is these spiritual beings that I wish to refer to as the spirit world and with whom we possibly share the universe. Certainly some will say that I mean to say the saints. This just brings us back to our earlier assumption that possibly all souls or spirits, with the exception of the angels, (who are in God's presence,) are in fact saints.

Fred Bonisch

It appears that using such terms as spirit world or spirit realm seems like such basic human expressions. Other than referring to heaven or the universe, we are somewhat limited in our description. Heaven seems to indicate the ultimate place of our journeying, the throne of God himself, and perhaps the true center of the universe from which everything originates. I like to believe that this is where the good spirits receive their reward by being able to live in the presence of God. Being in the presence of God may not necessarily be limited to a defined space since we consider that the entire universe is God's domain.

We have all heard of numerous accounts of near death experiences where those involved first encountered a bright light. One particular account describes the greetings of countless friendly spirits after having passed through the bright light. Although all of the account tellers returned to life and so were able to tell us about it, their experiences were usually life changing. They no longer feared death in the same way. Some even expressed their sadness at having to return to live out their earthly lives. We are not certain whether we are received directly into the place we refer to as heaven after our own passing. By recalling that Jesus assured one of the two criminals crucified next to him with the words, "today you will be with me in Paradise," surely gives us hope for a speedy entry.

We need to be careful when considering such concepts as they may well be viewed differently by many believers. When we give this greater attention, we find that much of this belief has actually been in practice by the Catholic Church since its formation. Many of its practices involve reliance upon the saints especially by seeking their intercession or help. Although we claim not to pray to the saints but rather to seek their intercession, by engaging in this practice, we have already set a precedent by assuming that they can be of assistance for our needs.

In our reliance on them, we have even come so far as to claim a saint for every one of our particular needs. For me it is St. Anthony who always seems to help me to find lost items. We even have saints who cover the whole gamut of our problems. Weekly town papers, nearly always contain thank you notes to St. Luke for a petition granted. Although I employ this practice to a lesser degree myself, I also realize that our thanks and praise should first and foremost be directed to God who allows all good things to occur. This should not imply that we should neglect the saints for their intercession but that we should be mindful of the hierarchical structure and the real source from which our graces come.

The practice of calling on the saints has been in use within the Catholic Church for nearly two thousand years. Protestant churches on the other hand do not believe in this practice. By this long held practice of engaging the saints, the Catholic Church acknowledges its belief in the existence of spirit beings aside from God Himself.

It is interesting to note that even Pope John Paul II credited Mary, the mother of Jesus, with sparing his life during the assault on him and so has done much since that time to venerate her. In fact, Mary is, for Catholics, by far the most venerated of saints. She has been placed above the saints and has become a main intercessor for Catholics in her special relationship with her son, Jesus. Catholics feel a special sense of closeness to her as she is viewed and venerated as a mother figure. Her so-called appearances or apparitions that have occurred around the world, have made her spiritual being visible to at least a selected few. Although not accepted by all, these apparitions are considered miraculous and continue to influence the Christian community. While it may not be common knowledge, Mary has been designated as the patron saint of our country.

We still find that many of our older churches are lined with the statues of saints, which further demonstrates the Church's belief in their ability to intercede on our behalf or even to bring about desired miracles. Churches are frequently named after the saints, a practice not only used by the Catholic Church but by other denominations as well. It needs to be pointed out that Protestant churches who do this will say that it is done to honor or recognize a deceased believer's work. They would rarely hold that their practice exemplifies trust in the departed.

This exemplifies how the two major Christian faith groups have very different viewpoints on this issue. While the Protestants give this little consideration, the Catholic Church has set aside November 1st and 2nd as special days for All Saints and All Souls Day. It has even declared November 1st as a day of obligation, requiring Catholics

Fred Bonisch

to attend Mass. It is interesting to note that these practices are not necessarily based on scripture but evolved through the Church's own teachings and past practices. My intent here is not to discredit this practice but simply to point out that the idea of saints, spirits and souls is by no means a new idea thought of by this author.

If in fact we accept that the saints are spirits who live in the presence of God and possibly have the ability to assist us, who then are these saints? We seem to have this narrow view, held mainly by Catholics, that saints are only those who have been canonized and recognized by the church. Webster defines saints simply as: "holy persons." The Bible refers to all believers as saints. (Romans 1:7, 8:27) (1Corinthians 1:2) Undoubtedly, we each have a different concept of how we view the qualities of a holy person. It would be hard to define when exactly we cross from being a regular to a holy person. Even when we attempt to judge ourselves, there are times when we feel a certain holiness while at other times we can hardly stand ourselves. In the end, I like to think that we are all holy people in the making.

The Catholic process of declaring someone a saint is known as canonization. The act itself is performed by the Pope and is usually done in a three step process. Anyone that is being considered for canonization, is obviously no longer alive. Frequently individuals are declared saints long after they have passed on. The qualifications required for such consideration can vary based on the individual's life-style and the circumstances under which they died. Martyrs, for example, who died for their faith are a special group and require no further confirmation. Others will be considered as a result of their exceptionally virtuous life. These also require proof of several miracles which have occurred through their intercession, a process that undergoes a great deal of scrutiny by the Vatican. Occasionally but rarely, the church even considers special circumstances under which a person may be canonized without the required proof of miracles. Mother Theresa of Calcutta and Pope John Paul II are currently being considered under this exemption. Since those considered for sainthood are always deceased, they themselves can no longer give an account of their lives or the miracles accredited to them. The church leaders believe that the church was given this power by Jesus himself when he told Peter that ..."*whatever you bind on earth will also be bound in heaven.*"

It is not the intention of this author to dispute the process by which someone is canonized. Rather, it is to point out that by the use of this selection process, the Church has again demonstrated its belief in spiritual existence.

With the exception of the martyrs, each circumstance for canonization may be unique. An example of this is Monsignor Nelson Henry Baker 1841-1936 of Buffalo, N.Y., former Pastor and founder of Our Lady Of Victory Basilica in Lackawanna, N.Y. Thus far the first step towards his sanctification known as Servant of God has been completed. Because of his extraordinary service to his community and church by creating an orphanage, a home for infants, a facility for unwed mothers, a maternity hospital, and for building one of the most beautiful Basilicas in the U.S., it was felt that he is deserving of this highest consideration. He was laid to rest in the adjacent cemetery of the Basilica after his passing in 1936. In 1999 his body was exhumed and laid to rest inside the Basilica. During the process of moving his remains, the workers found a small vault which contained three vials of his body fluids. One of the three vials contained Fr. Baker's blood. What was astonishing about this find was that the blood appeared as fresh as when it was placed into the vial some sixty-three years earlier. The blood was examined by local medical experts, as well as by the Catholic authorities in Rome, and the finding was authenticated. This unexplainable occurrence has, of course, added special consideration in his process of being declared a saint. Confirmation of miracles that are accredited to his intercession are still required for the final approval.

As we look back on the period of five hundred years prior to Vatican II, however, we find that it was usually only clergy or religious who were found worthy of this consideration. It may cause us to question this imbalance and wonder about the Church's view of holiness. Is being either clergy or religious member a prerequisite for holiness? How do we compare the parents who have raised ten children with someone who has lived in solitude for much of his or her life? It seems that we can only pick and choose without the ability to look into someone's soul; we then realize that many true saints will never be recognized for their virtues.

Recently while having lunch at a shopping mall food court, I observed a young woman having lunch with her three young children just two tables from mine. Besides the two little girls of about five and three, there was also a boy of about eight or nine years-old.

The boy was confined to a special wheelchair and was held in place by several straps. What was most amazing was the patience this young mother displayed during the process of feeding her handicapped son while at the same time responding to the chatter of the little girls. Her face expressed a warm gentleness and I sensed the love she projected while feeding her helpless son; for me, there was absolutely no doubt that I was watching a saint in action. The realization that this would probably be a life-long requirement of this young and loving mother left me in awe and admiration for her.

We can see that parenting, the most important commitment for the continuation of mankind, is not considered a virtue by itself sufficient for canonization. What truly constitutes holiness and who comes under its umbrella, will ultimately be decided by God alone. He is the one who truly knows our lives and our struggles and so can pass a fair and honest judgment on each of us.

I doubt that any of us would consider ourselves as saintly material. We are usually hardest on ourselves. Yet others often see something special in us that we ourselves overlook or do not even consider as unusual or noteworthy. It may just be something as simple as being a good listener or having the ability to acknowledge others in a respectful manner. It should make us realize that in spite of all our shortcomings, there is that hope of saintly moments that show themselves at least now and then.

We should begin to recognize that part of the Christian world at least accepts the existence of saints as spirit beings. We have also learned that the Bible refers to all believers as saints and this includes the living and those already departed. Since, for the purpose of this book, we are interested whether unseen forces in the universe can assist us, we shall concentrate on the souls of the departed spirit-saints. For our purpose, let us assume that these souls of spirit-saints now exist in the presence of God. I could easily assume and am in fact hopeful that God in his mercy does not make that distinction and that all souls who come into His presence are considered saints.

So far we have talked about our souls, the angels, the spirit-saints and briefly the demonic spirits. Without attempting to explore these entities further, we recognize that the church makes a distinction between the angels and what we may refer to as the saintly spirits. Angels were created to serve God and as such continue to be angels. Saintly-spirits who also live in the presence of God are of a different nature and so may possibly serve Him and assist mankind on His behalf. Although both groups give praise and glory to God, they may serve Him in very different ways.

Our Perception of Heaven and Hell

Heaven and Hell are two concepts that we often use very freely without giving much thought to them. Yet, they are the future dwelling places of our souls. Depending on the choices of this life, we will spend an eternity in either one or the other. These places represent the good and the evil with God on one side and Satan on the other. Both are unseen spirit forces that hope to draw us to them but allow us to choose for ourselves. When we consider any spirit being, regardless of what we may decide to name it, we need to make a clear distinction between those who dwell in Heaven and those who are condemned to Hell. Other than acknowledging the existence of Hell as a separate and confined place in the universe, future references to the spirits in any form shall refer to those whom we believe live in the presence of God.

Heaven for us can have several meanings. Looking up at the sky, we often refer to it as looking up at heaven. Simple sayings such as "the heavens opened up and poured out buckets of water" are used freely without serious thought. We frequently say that we want to go to Heaven. Heaven is where God resides. It becomes more serious when someone passes on and we say that he or she is now in Heaven. If they were at all special to us, then that is the place we want them to be. As humans we think of it as a well defined place with walls and possibly a large gate with St. Peter as its keeper. Since we can't define Heaven or its location, we have to assume that it too is part of or within the universe. No human being, no matter how holy, has been able to define

Heaven. From the Book of Revelation, (Revelation 4:1-11) we have John's brief account of his vision of Heaven. Even that is somewhat cloudy since it is unclear whether he saw this in a dream or in reality. More specifically, even his account gives us no indication of its location or whereabouts.

Jesus has referred to this place we call Heaven as the place where God His Father resides. During His suffering on the cross, He was assuring one of the two criminals who were crucified with Him (Luke 23:43) that *".... I say to you, today you will be with Me in Paradise."* At another point while talking to his disciples (John 14:2) He tells them, *"In My Father's house are many mansions; if it were not so, I would have told you. I go to prepare a place for you."* This should assure us that there is such a place as Heaven, it will be both beautiful and spacious and we will be welcomed there.

Since we are left with the assumption that Heaven is somewhere in the universe, it could well be that Heaven is not just a defined space but that the entire universe itself is at God's disposal. If all of this is His Kingdom, why then would He have a need to select a special place to be called Heaven? Although I have never seen or heard this mentioned, perhaps Heaven is not a place at all but a state of spiritual perception. Spiritual Teacher Eckhart Tolle in his book "A New Earth Awakening to your Life's Purposes" states: *"We need to understand here that heaven is not a location but refers to the inner realm of consciousness"*. While we can consider such abstract views, we usually prefer to relate to solids and objects and so we each have our own perception of what heaven may be like.

If we consider the universe to be all-inclusive, then Webster's definition as "the totality of all things that exist" is probably the best description. For us this is simply an overwhelming concept. Being used to finite concepts makes dealing with infinite ideas such as the totality of all things nearly impossible to comprehend. If it encompasses all things that exist, it should then include heaven and hell even though it is outside our understanding. It simply includes everything that God has created and so we can comfortably refer to the universe as all of God's creation.

During a recent discussion with Christian friends about the existence of a spirit world, I was pleasantly surprised by the fact that the majority of the group agreed about the spirit's unseen presence and their willingness to help us. Someone, in fact, expressed that the good spirits whom we assume to live in heaven or their assigned space, take pleasure in helping us. On the other hand those who are condemned to Hell are prevented from helping us, and are, therefore, grieved or saddened by this fact. It would seem understandable, from our human perspective at least, that the inability to help your needy loved ones on Earth could become a sheer hell. I wish to remind the reader that these were simple exchanges of thoughts and are not intended as proof of spiritual abilities.

Catholic theology teaches that the dead who die in Christ live in Christ. If the deceased supported us on earth then they must also continue to do so in their resurrection. If we accept that the spiritual world is there to help us, it may be of interest to consider their abilities or powers. Are they mere observers without any special ability to act, or can they in fact, perform what we may refer to as miracles? Again we can only speculate about this ability. This becomes especially confusing when we think of the many disasters and personal hardships that each of us has experienced. Why are we helped in certain instances while at other times we feel so abandoned that we even question the existence of God himself? This is certainly a question for which we would like to have an answer. In the business world, usually only the top level managers know the true situation of their operation and so decide the best course of action, often to the worker's dismay or understanding. I trust that it is so with God as well who alone knows the overall plan for our life on earth and for our eternity to follow.

Since at least part of the Christian community already believes that the spiritual world or the saints can assist us with our needs, whether that occurs through intercession or by direct means, I wish to follow this positive assumption. Whether we call it the spiritual world, the saints or the universe, I would like to think of it as the source. Since, however, we usually think of the universe as the entity which makes up the space in which all beings exist, perhaps this description then will serve us best. We can compare the universe in the way we define the church. We don't refer to the congregation, the buildings, and the clergy separately; instead, all of them together make up our concept of the church. So too the universe represents all that exists, including all of us, the living and the dead, as well as Heaven and Hell.

Our Contact with the Questionable Spirits

Curiosity about what lies beyond our visual realm has been part of every generation. Although we have learned much about our solar system within the universe, the spiritual aspect, however, continues to be a mystery. It has baffled generations to this very day. We may have overcome many outdated superstitions, but it remains a mystery nonetheless. Thoughts about spiritual beings still put fear and superstition in people's minds and yet, in spite of it, we have never lost the curiosity to know and to have a peek inside this unknown but intriguing world. The entertainment industry continues to provide us with new ideas and grotesque visions and so the idea of a spiritual place is often looked upon with mixed perceptions.

How often have we heard of the warning not to play with Ouija boards or Tarot cards or to consult fortune tellers, as this will make us vulnerable and leave us exposed to allow demonic spirits into our lives? I do not attempt to prove or disprove this notion, however, I recall playing with a Ouija board as a group and fun activity just one time. The answers we received to our questions were so precise. What we were told would happen shortly then came about so quickly that we discarded this game immediately.

It is interesting to note that even the medieval church believed in the power of spiritual beings as can be seen from the book: *Malleus maleficarum* (English title *The Hammer of Witches*) which was written by two German Dominican Friars and Inquisitors, Jacob Sprenger and Heinrich Institoris in 1487. Instead of spirits, however, it referred to the Devil and demons that would take control of human beings, mainly women, and so turn them into witches. In spite of the fact that the Catholic Church banned the book in 1590, the concept found such an acceptance that *Malleus maleficarum* became the guide book for witch trials throughout the courts of Europe for some two hundred years.

Conservative estimates claim that probably 40,000 to 100,000 mainly innocent women and girls were executed as a result of this extreme belief. This dark perception was also held in some of the earlier settlements in the U.S. and was perhaps most notoriously practiced by and known as the Salem witch trials. It can be argued that only a few people lost their lives on our soil through this strange belief, but it occurred nonetheless.

It is interesting to note that even today, there are women who proudly and openly claim to be witches who associate themselves with the worship of Voodoo, Satanism, Vampirism, Wicca or Santeria. This is not only limited to women but men who claim to be part of these cults as well. We can see that although we live in the twenty-first century, certain people are still drawn to what most of us perceive as the dark side of the spirit world.

From the past we can see that the influence of spiritual beings was viewed with some degree of fear and so projected a more negative force. Even today we have not totally overcome this thinking as it is reiterated to us in books and films. People still consult psychic mediums, also known as spirit messengers, often for the purpose of contacting dearly departed loved ones or relatives during séances or simply to learn what the future may hold for them. The mediums or spirit messengers claim to have a special ability to communicate with the spirits of the departed that usually surround or are assumed to be near the person who is hoping to receive the spirit's feedback. From observing the performance of some of the well known psychic mediums, we find that they are often surprisingly correct in the statements that they pass on to surviving relatives or friends. Their ability to describe the spirit person with whom they are purportedly communicating often adds to their credibility and leaves the person seeking the information in awe. We find that, although it is not often expressed, that there is something in us that wants to believe in the existence of spiritual beings. Perhaps it is an unrealized tendency to project our own immortality in this manner.

Unquestionably, many people believe that certain individuals have the ability to communicate with the spirit world, something I wish to neither reject nor defend. We need to be careful since many so-claimed mediums are fakes and in it for profit. There are, however, the exceptions with special and honest intentions that will, of course, always cause us to think and wonder. What makes this issue so interesting is the fact that it is and will likely remain a mystery. It seems certain that we are not meant to know all these secrets. If on the other hand some of the mediums are telling the truth, it may well serve as a reminder to us that we are not alone. It is probably fair to assume that all of us have had experiences that were so profound that they sparked our curiosity and caused us to ask, "Who is out there looking out for us?"

My own journey and the experiences I was allowed to enjoy have convinced me that I was truly graced in so many ways while facing the obstacles that were before me. It was not just one isolated incident but the culmination of numerous such events that has brought me to the belief of supernatural interventions in my life. Although other things encouraged me to write this book as well, it was the experiences of these special events that became the prime motivation for doing so.

Unusual Occurrences in Our Lives

In my first book, *Children Of Our Own War, A Boy's Journey*, I described my family's experiences while I was growing up during World War II and the postwar years in Germany. Since my book only described what we experienced during this dark period in history up to about age fifteen, several of my readers have asked whether I plan to write a sequel describing the events of the following years. Although I felt flattered by their request, I didn't consider my life all that interesting, at least not to others. In its place I have chosen to share incidents of my life that are not meant to draw attention to myself but rather point to the extraordinary nature and circumstances under which they have occurred.

We all have unexplained happenings and so I am no more at an advantage or disadvantage than anyone who will take time to read this book. Although I started this project several times before, I stopped writing out of concern that some of the events described are too personal and should remain so. In the end, however, I have come to realize that by sharing my experiences, others too may begin to take notice and so become more in tune with the unusual and often miraculous occurrences in their lives.

The concept of this book may be easier to accept for those who believe in a higher power, namely God, but it is by no means a prerequisite. Nor is it intended to claim that any religion has an edge over any other. We are taught and believe that we are all children of the same God and so it is not for us to draw conclusions of superiority or of being in a select group.

The reader may notice my Christian influence as he/she continues reading this book. I mention this with all due respect to other religions or beliefs. I do not claim to be perfect and admit that I am far from being a saint. I have made my share of blunders and will not escape them in the future. Yet, in spite of our imperfections, we encounter blessings that often come out of nowhere and as a total surprise. When we look back on our lives, we come to realize the tribulations and the miracles we have encountered along the way. There are those who may even ask themselves, "How did I get to this point on my own?" This then again becomes the real question we need to ask ourselves, "are we alone in this journey of life?"

Recognizing Our Blessings

While I was growing up during World War II and the postwar years in northern Germany, we had few expectations besides staying alive. Since our father was drafted into the German army and became a prisoner of war, Mom was left to take care of four young children during very difficult and often frightening times. With the war going on around us and while trying to exist on starvation style rations, survival was foremost on everybody's mind. This often hopeless period of my early childhood has naturally influenced much of my adult life in that I realized the importance of having family and that life is worth living in spite of the many obstacles we encounter. Yes, we will experience dark periods, but the sun will eventually reappear from behind the clouds.

Through the grace of God, my family survived this dark wartime period in history and even Dad returned home after having spent seven years as a prisoner of war. This in itself was miraculous considering the many others who had lost so much. In retrospect, I realize that God had an unusual but challenging and interesting plan for my life. This plan was to include leaving my homeland and having the opportunity to travel extensively. It was through this activity that I encountered many good people, especially in Asia, who did not know or believe in God as we do. They demonstrated such respect, dignity, and trust within their society that I began to wonder about how much our religions really influence our western morality for the better. Yet in spite of this impression, I feel that our belief in the existence of God, regardless of the religion, gives us a purpose, comfort, hope, and a direction. Life no longer becomes just a here and now, take what you can get, enjoy it while you can. Through our belief we become connected with something that may be unknown to us but is a power far greater than our human mind can even begin to comprehend.

Nothing in this universe happens without God's knowledge and approval. Besides the gift of His son, Jesus, He gave us the ability to choose, knowing how unreliable we can be. It may even be debatable whether this freedom serves us as a blessing or a curse but in spite of it God wanted us to have this ability. As parents we can understand what it means to let go of our maturing children by allowing them to select their own lives while realizing that they will make both good and poor decisions. At the same time we stand ready to advise or help them to correct or overcome their mistakes. These are simple acts of unselfish love we feel for our children and we gladly sacrifice our own comfort in the process should it become necessary.

I want to be careful so as not to turn this into preaching or a Sunday morning sermon but we need to understand that God too stands ready to help and assist us. Although we usually pray to Him, we can't say for certain whether His response always comes from Him directly or whether He allocates the response to some other form of spiritual beings.

From childhood on we have been taught that we each have a guardian angel to protect and to watch over us. Yet unfortunate things happen to children and adults in spite of this belief and so we must wonder whether watching over us was really their intended purpose.

From our limited human perspective it is difficult, if not impossible, to conceive how one God can monitor the prayers, wishes, hopes, lives, and thoughts of the six billion living people, not to mention the many more souls who have already returned to Him. To be able to do all of this in addition to controlling the universe seems not only overwhelming, but incomprehensible, almost unbelievable. Yet this is what we are taught and believe. My intent is not to dispute this ability but to point out what wonderful support the saints or the spirits are in serving and supporting God in this enormous effort. For the religious of my own faith, some of whom may disagree with me, I wish to point out our relationship to Mary, the mother of our Lord Jesus, and to the saints. By petitioning them for our needs, we assume that they already have this relationship with God.

In the New Testament, Jesus assures us that: *"whatever you ask the Father in My name He will give you."* (John 16: 23) It is something that I think many of us are struggling with, for who of us does not have unanswered prayers. This issue frequently arises during discussions with other Christian friends. For lack of a satisfying answer, we usually determine that perhaps the reply "No" becomes the answer because God knows what is best for us in any given situation. From the reaction of the group members, however, not everybody seems ready to accept this as the proper justification.

In spite of the fact that I consider myself an active Christian, I want to rely first and foremost on myself rather than allow God to control my life. Making God truly a part of our lives and including Him in all of our daily trials and decisions takes real practice and commitment. Although I admire those who live by this practice, I still lack the commitment to truly consider myself their equal.

We all have unexplained phenomenon whether we are believers or not. This then becomes the real issue of this book. Many have experienced these unusual events even without asking for them and yet they seem so appropriate at that moment. We need to learn to recognize and acknowledge them for what they are. We may call them luck, coincidence, miracles or gifts but by becoming cognizant of them when they occur, we begin to sense the greater power that is behind them. This book points out several of such unexplained occurrences in my own life in the following chapters. Some have left me in awe while at other times they occurred almost unnoticed. It is through these events, that I have come to realize that some unseen being or beings are truly looking out for me and my family.

It would serve us to recognize and accept that these unusual phenomena are simply without limits in their occurrences. Even when we realize that something special has occurred, we often fail to recognize the power behind them. We have to accept that we will be able neither to prove nor disprove the intervention of such forces as a spirit world serving God's will. Having said this, however, I take the side of the believers. Too much in my life has occurred that is just beyond normal reasoning to be viewed as luck or coincidence. By sharing these experiences, I hope to inspire the reader to at least consider the possibility of such external interventions and to become cognizant of them in their own lives. Trusting that we have that invisible assistance can be a comfort and should in no way distract us from our religious beliefs.

Although we can debate this issue to no end, by sharing my own experiences, I hope to elevate this to a greater awareness. Perhaps it will at least inspire the reader to become open minded enough to consider such possibilities. The first of my examples may not be the most convincing, however, not only did it assist me during a critical time, but more importantly it was my final inspiration to write this book. In fact, our inspirations are perhaps the most unrecognized and overlooked gifts or supports that we receive from our unseen partners.

As any author will attest, the gift of inspiration becomes the motivation to write and to create. Even while writing this book, I feel myself drawn to doing it. Here too, I feel the spirits energizing me and calling me to do this work. Although it will give me the ultimate satisfaction after its completion, it will hopefully serve others to come in touch with their own calling and so I see this as the purpose as well. None of this simply happens by coincidence, but we all have specific functions to which we are called. The fact that they are different for each of us does not make us better or worse but just different and unique. We are part of the universe in which we have certain roles and duties to fulfill.

We may think of ourselves as insignificant, but our actions, no matter how small they may appear to us, are precisely calculated to meet the perfection of the universe. Whether we call these insights, inspiration, or intuition is less important, instead, they may well be our real calling. When we begin to pay attention to these often subtle inspirations, we may come to realize their possible purpose and even discover our hidden talents in the process. Since, of course, we also receive negative inspiration which we attribute to the dark forces in the spirit world, it is important to discern the intent. Through this awareness, we may begin an important step in our partnership with the so-called spirits. The sharing of the following event may well be an example of how diverse inspirations can affect us.

Fred Bonisch

Some time ago I was faced with an important financial decision. While standing with my back leaning against the kitchen counter, I prayed for guidance to help me make the proper decision. While I was hoping to receive some form of response that would help me decide, something encouraged me to pick up a neglected fortune cookie that had been laying on the kitchen counter. A fortune cookie was not the answer I had hoped for and so I was ready to quickly dismiss this thought. I had read many fortune cookies over the years without ever considering them as anything other than fun. Was I again expecting a specific answer, something I had hoped for so many times before? While hesitating and waiting for something better to occur, I kept staring at the fortune cookie and wondering what it could possibly tell me. Finally, my curiosity led me to take hold of it and I suddenly felt an anxiety to learn its message. After breaking the oddly shaped cookie, the little white note dropped onto the counter. As I read the small two-line note, I was astonished by the message: *"Trust your intuition. The universe is guiding your life."*

My immediate reaction was simply one of astonishment. What a perfect message this was. Not only did it give me an answer to my prayer but it gave me a choice with an assurance that I would be correct regardless of what decision I would make. The message instilled a confidence that I never thought to question. It seemed as if a voice kept reassuring me that I would make the correct decision. Rarely have I felt so peaceful about what I considered a serious choice. My anxiety declined in seconds, all my doubts vanished and I made my decision happily. Although I had no doubts, it was not until several days later that I learned of the positive result of my earlier judgment.

I have taped this little note to my computer as a reminder and have applied its message numerous times since. In fact, trusting that the universe is guiding my life has become a great comfort and reassurance to me. However, many good and honorable believers may claim that I have tried to bypass God in the process. In fact I recall the words of a radio Bible teacher who claimed that such responses are never from God. In reply I would argue that my prayers are always directed towards Him and His Son, Jesus. With respect to those who make this claim I would ask what good are our prayers if we reject the response we receive as something not coming from God? While I agree that

The Universe, is it Guiding Our Lives?

we need to be aware of the devious means employed by the Devil and his helping spirits, we can not allow ourselves to become so paranoid that it overshadows our trust in the goodness of God. I realize that the occasional response we receive to our prayers may not come in the form we often expect. This, however, should not cause us to automatically dismiss it as the work of the Devil and his spiritual helpers.

Much of my life is filled with other unusual but real events. Many of them have truly been miracles for me and for my family. I too have experienced painful and worrisome periods in my life and yet through prayer and trust, I have been able to overcome them.

While I can not claim to have special graces or powers, I have been blessed many times. By sharing them openly, I hope to bring honor to God Himself and recognition to whatever good spirits He has assigned to help me.

A Miracle of Healing

The next unusual and puzzling event that I wish to talk about dates back to when I was about the age of one. I learned of this experience from my mother when I was old enough to understand. By the time I was one year old, World War II had already began and much of Germany's resources were being allocated to the war effort. Dad had been drafted into the German army the previous year and so Mom was left to take care of my five year-old brother and myself. As is frequently the case with young children, they'll experience allergic reactions to food, liquids, animals, or simply the air around them. I had developed a severe case of head sores, also referred to as cradle cap. According to Mom's description, my entire cranial cap was one large patch similar to that of a severe case of eczema that would bleed and then turn into dried blood. The town nurse, who would usually act as substitute for the town doctor, came to visit us twice daily to wash this dried cap from my head. Needless to say, it was a very painful and bloody event that required a change of bed sheets several times each day. While telling me of this dilemma, Mom would shake her head as she recalled my screaming each time the nurse came to visit. In an effort to heal me, we visited several doctors and specialists during the course of a year without improvement.

Because nothing seemed to help, I began a treatment at our local doctor's office where I was required to lie under a special lamp for a given period of time every other day. After weeks and numerous visits it became clear, however, that this too showed no signs of helping and so my painful condition remained unchanged. While listening to Mom telling me of this period, it was obvious from her expressions that it was a stressful period for her as well. After I had suffered with this condition for nearly a full year, the doctors termed my situation as problematic but failed to have the answers to cure me. This left Mom with no hope for an improvement or healing from the medical community. In desperation she began to ponder about alternatives.

Our north German town had retained much of its medieval appearance, where legends and myths had been passed on from previous generations. It was not uncommon for some people to accept old rituals and traditions. One of these was the ability by several of the older woman to perform somewhat mysterious deeds. Among these was the ability to foretell a person's future and to perform certain types of healing. The healing of warts was one such cure and was often performed when traditional methods had failed. Although no medication was used in the process, warts that had often plagued a person for months or years would usually be gone within just days. Once such ladies became too old to perform these rituals, they would then pass this special ability on to younger women of their choosing.

My family was one of only a few Catholic families in an all-Protestant town. To consider the services of these specially gifted ladies was frowned upon by the clergy and fellow Catholics and probably by many of the Protestants as well. Yet often out of desperation, people would engage the services of these ladies in hopes of relieving various concerns. With Mom realizing that she could no longer hope for healing from the medical community and wishing for me to finally become well, she also decided to consider the service of one of these ladies. Anyone who had used their special powers for whatever purpose would usually talk about it in a whisper, as if to pass on a special secret. Naturally this only added to the mystery surrounding these services. Since these ladies were well known within the community, it was usually not difficult to arrange for a meeting with them. The actual ceremony was conducted at either's home.

I am certain that it was hard for Mom to seek the ladies' special services, and even now she only reluctantly talks about the event. Once she had arranged for me to visit one of these ladies, I was taken to her home for the healing ritual. The lady placed her hands over my head and whatever was said was said in silence. Upon completing her silent ceremony, she instructed Mom to bring me back for one more healing ritual. The second visit was a repeat of what she had done during our earlier visit. She assured Mom that nothing further needed to be done and sent us home.

Amazingly, within a short time following this mysterious exercise, I began to experience a complete recovery and soon grew a full head of blond hair. Since the medical staff had given up any hope of healing me months earlier, I trust that my healing occurred through the intercession of a supernatural power. Although the extraordinary ability of these ladies remained a mystery, their results, however, could hardly be questioned. Whenever I think about and question the source of this healing event in my life, I recall the words of Jesus who told us that, "*A good tree cannot bear bad fruit, nor can a bad tree bear good fruit.*" (Matthew 7:18) So, I trust and believe that it was through the grace of God and perhaps through the intercession of His good spirits that I was healed in such a miraculous manner. Some Bible scholars will probably argue this assumption by pointing out (Exodus 7:10-12), in which Pharaoh called upon his sorcerers to mimic the work of God, but eventually their works were swallowed up. Since it can be argued either way, I would rather believe in the power of God since obviously He had a plan for me that I was destined to fulfill.

Listening to Our Inner Voice

    I have often wondered whether we come into this world with a destiny. It is unique when we think about the fact that we are all created to be different. When we take into account all the people who have gone through life before us and are currently with us, we have to admit, our destiny is truly one of the great mysteries and wonders. Although we may inherit our general looks and certain personality traits, we are still unique. As children we require the nurturing of our parents but as young adults we already have the desire to go our own way and make our own decisions. Yet not fully mature and still lacking many of the basic life experiences, we often make poor decisions and as a result, we begin to question our own abilities.

    Mistakes are usually looked upon as negative experiences. Although we prefer to avoid them, they are often the greatest influence in building character and maturity. If we hope to deal with difficult situations or try to understand others in such circumstances, we need to experience them first hand, as painful as they may be at the time they occur. Even here I trust that the universe can help us to learn life's valuable lessons and set us on course in accordance with our destiny.

    During their high school years, all three of my children worked at local restaurants doing various jobs. Although it earned them their own spending money and so gave them a certain independence, they also realized that it was not something they desired to do as a life long career. Experiences, both the good and the more troubling, help us towards our maturity and eventually determine our course. Although this should at least be the general format for success, we realize that many young adults get detoured by indolence and the search for instant gratification and quick and easy money. Often times the greatest gift we receive, namely our freedom to choose, becomes our own detriment. This applies to both young and fully mature adults alike.

If we trust that God may have assigned spirit beings to assist us, then we need to listen to our inner voice, our "gut feeling," for the direction we are intended to go. Learning to listen to our inner voice can become valuable guidance. By suppressing this important voice, however, we may well shut out any direction and help that is being offered to us. In fact, we reject and block the positive forces from working for us when we insist on doing everything ourselves.

I too have attempted to take directions that in hindsight would have turned out as dead ends. Even the spirits must respect the free will given to us by God and so our acceptance or rejection of their gifts can become a blessing or a curse for us. When we recognize that we have taken a wrong turn or made a bad decision, we can change and attempt to correct our mistake. This may well become the moment in life when we are most open to receive that often unrecognized support.

One of our negative human traits is that often when we ask for help, we already have a certain expectation of how this help will present itself. I must admit occasionally being guilty of this trap myself and so I will fail to recognize the answer or help if it appears to be different from my expectation. Worse yet, I have attempted to negotiate with God, the Creator of the universe, for what in retrospect appear to be somewhat petty requests but which seemed rather important at the time they were made.

Coming back then to our question whether we each have a destination for which we are predestined, I like to think we do. Whether we ever reach what we have been destined for probably depends much on ourselves. It is important to realize that we have different courses laid out for us and so success becomes something very different for each of us. It may well be that we fall short of our own goals, which is naturally disappointing. If, however, our destiny has allowed us to prepare our children for their future, then we can consider ourselves successful after all. We need only to think of today's college tuition

The Universe, is it Guiding Our Lives?

costs for our children to understand that this view may not be so far-fetched. As parents it is our hope and joy for our children to have a good or better life. This then becomes a goal and perhaps our destiny is to prepare each new generation for what lies ahead.

One of the great tragedies of our time is that we often measure success (or whether we achieved our goals) by the amount of wealth or prestige we have gained. True success should really be how we view ourselves. Do we respect ourselves or are there things that pull us down and prevent us from feeling satisfied with who we are? Many, if not most of us, don't come to this realization because of the business of life; often the deathbed finally becomes the place of reflection. Even here God will hear of and forgive us our trespasses if we simply ask for forgiveness.

Someday we will become part of the spirit world and recognize how insignificant our concerns have been, and how much we have subjected ourselves to unnecessary worrying. How little we have recognized the good forces in our lives when we looked at them only as our own successes. By coming to this realization, we should take notice of what the universe has to offer while we are still here on earth.

Desires Becoming Reality

The war and the postwar years were hard for Mom. By 1943 our family had grown to three boys and one girl. Dad had already served in the army for nearly four years and the entire burden of caring for our young family fell onto Mom's small shoulders. 1943 was also the year we received the unfortunate news that Dad had become a prisoner of war. Although it was sad news for our family, at least we knew that he was still alive. Many men from our town had been either killed or declared missing in action. Recalling much of this fearful and difficult period, I was prompted to dedicate my earlier book: *Children of Our Own War* to our mother and all the other mothers of the war.

During the final years of World War II relatives who had lost their homes as a result of the frequent bombings in a nearby city came to live with us. Other relatives who were driven out of the Russian occupied German territories in the east came to stay with us as well and so our small house became filled with people. We were a blessing for them and their stay with us during this frightening period became one of my most cherished childhood memories. Even though we were sharing the miseries of the war, it was in fact this sharing that made this period so memorable and helped all of us to survive.

By 1945 the war was finally over and we were fortunate to be occupied by American forces. This period would greatly affect my life in a positive way, something I would come to realize only years later. Dad remained a prisoner of war for an additional four years and was finally able to come home in 1949 when I was nearly ten-years old. In spite of the hard times, we had survived the bombings and the food shortages of the war. With Dad finally being home again, we were a family once again and so we felt truly blessed.

We have a tendency to take our lives for granted, but as I look back I realize how miraculously we had survived this dark period, and that in the end we still had each other. When I recall the many who had lost fathers or brothers or all their possessions, I realize how blessed we were to have survived in spite of all that we had experienced during those war-time years.

The occupation following the war was a time that affected me in a positive way that would first show itself several years later. As a young boy I was fascinated by the movements of the American military personnel. American soldiers moved about with a certain casualness that Germans took notice of and that somehow impressed me personally. Also, the military seemed to lack nothing in their provisions and often shared chocolate or even sandwiches with us. I realized even back then that someday I would want to be part of this lifestyle but also recognized that it would have to wait.

From watching the constant postwar military movements by the occupation troops, I particularly recall my boyish fascination about owning a Jeep someday (even though I was only about seven or eight years old at the time). Although it seems like something insignificant now, it was something that often kept me awake at night while I was mentally cruising through the countryside in my Jeep. I mention this childhood fantasy because it remained in my memory and was to become a reality later in my life.

I became aware of my desire to travel at an early age and even a bus ride into the city was an exciting experience. I recall shortly after the war, Mom and I had to travel by train to the nearby city of Hildesheim where I was to undergo a hernia operation. These train rides, only one hour in duration, would remain for me an unforgettable experience. Although I had this strong desire to travel even at an early age, neither of my brothers nor my sister shared this desire to the same degree.

Mention of these recollections may appear insignificant to others, but it seems almost miraculous how these things would later become a reality in my life. It is almost as if someone began to lay out my life, waiting for the proper time to bring these desires to fruition. I lived through those early years much as most other children. I was fairly outgoing with others and friends.

I finished school and then learned a trade as a machinist. This choice was not necessarily out of interest or desire but one was expected to learn a trade to prepare for the real life ahead. This three and a half year apprentice program would become one of the hardest periods of my young life. Even after all these years, it continues to haunt me with an occasional nightmare. Mistreatment of trainees, which often included beating or kicking, was still considered acceptable by society but was a cause of constant worry and fear for me. Having lived through this troubling period, however, I was prepared for other difficult situations that I would encounter later in life.

My big break came within a year of completing my apprenticeship program. I was about to turn twenty when the opportunity to emigrate to America presented itself. Much to my parents' dismay, I immediately decided to go. I had a trade and was eager to experience the American dream; even though, I didn't quite realize the full consistency of that dream. It was a chance I was willing to take. I recalled that much of my desire to come to America was already set in motion during the time of the occupation back in 1945. With my excitement about traveling and leaving my home and family for a new adventure in a strange country, it seemed as if I was predestined for this serious new step in my life.

Coming to America

    Change and the unknown have always been worrisome to many, if not most, people. I recall my parents wondering how I would get along in America, realizing that I was not even able to speak English. They wondered what I would do and how I would support myself. I am sure that there were many other questions that plagued them. I refused to concern myself with these issues; perhaps it was a bit selfish on my part. Even my brothers and sister occasionally asked whether I was afraid to go so far into something unknown. For me, knowing that I was going to America was all that mattered. The long fourteen hour flight to New York was going to be the ultimate of my limited travel adventures. Experiencing a new country, language, people and opportunities was all that occupied my mind.

    It wasn't really until much later in life that I began to realize how different I was from my other family members by my unusual desire to travel without that fear of "what if." It was, in fact, this energy combined with my early desire to come to this country that would finally make it a reality. When ultimately a sponsor offered to help me, it was as if my dream was falling into place. Was all of this just simply circumstantial? I was still too young to question this aspect then and just credited it to my good fortune.

    Within a week after my arrival in the US, I began working in a local dairy making cheese. It wasn't Wall Street but it helped to pay the bills. This job, however, lasted for only six weeks and then I found myself unemployed. I still was unable to hold a conversation in English and so my future was in doubt. Someone again was watching over me, when two weeks later I was hired by a local aerospace firm to work as a machinist. I was told that German trained machinists had a good reputation for their work ethics and skills and so the lack of my language skill was overlooked. I was placed into a department whose manager had also emigrated from Europe and so he took me under his care.

It was simply amazing to see how well I was being guided and taken care of in my new life. Again I counted myself lucky. While working the night shift, I attended the local high school during the daytime and within a few months I became comfortable speaking and understanding what would be my new language. Although I didn't realize it at the time, my newly found employment would become a true blessing and a steady workplace in the years to come.

Another Dream Fulfilled

As a young man arriving in a new country there were things I wanted to do and to have. A car was, of course, very high on my list and once I had obtained my learner's permit, I purchased a 1955 Mercury hardtop with fender skirts. Owning my first big American car was truly living the dream and I quickly sent pictures back to my family in Germany boasting about my new life.

As a young boy growing up in Germany, I had often fantasized about the possibility of flying a plane. Even back then it was considered an elitist sport, something very few people could afford. I was still a young boy and so the possibility of flying a plane myself seemed simply unimaginable. Occasionally a small, single-engine plane would appear and land on a nearby grassy field. I recall those occasions when I would run about a mile to reach the pasture-like landing field. Just seeing that small plane sitting there, reminding me of an oversize bird, was always an exciting event. Being allowed to come near the plane was exhilarating and I would usually stay until it took off again. This affected me so much that I frequently experienced my desire to fly in my dreams at night. Although it was only a dream, it remained with me as a wonderful memory and caused me to wonder what being able to fly a plane must be like.

It was only years later that I would be able to make sense out of all of this. The urge to fly kept nagging me and so, within two years after arriving in the U.S., I decided to make it a reality. I began to take flying lessons at a local airport. I recall my first solo flight as one of the proudest moments in my life. Within the same year I earned my private pilot's license and was qualified as an official flier. I truly enjoyed flying and always look back on it as one of my most enjoyable achievements. It was not until much later that I would come to realize that my small accomplishment had another, perhaps more important purpose.

Here again was something instilled in me that I was destined to accomplish. Although I did not pursue flying as a career, the dream, however, was to be fulfilled through my son Marc who would later become an airline pilot. Earlier it was pointed out that sometimes one of our accomplishments is to prepare the next generation to bring a destiny to its completion. I have often asked myself whether this was again just another coincidence; one more of my childhood dreams had come to fruition. Again I marveled at the higher power at work in my life. Whether it was directed from God Himself or through His wonderful designated spirits will, of course, remain a mystery. The real miracle lies in the unique sequence of events which will become clearer in one of the later chapters.

A True Act of Kindness

It is easy for us to be dazzled by the more obvious and extraordinary events. Unfortunately, we often overlook the many smaller or seemingly inconsequential daily events we all experience. Each one of them is special, set apart only by what we perceive as their magnitude. For example, how do we compare the avoidance of a last-second automobile collision with winning in the lottery? Both can have profound but very different impacts. How often have we driven from point A to point B while being mentally preoccupied with other thoughts only to wonder how we arrived at our destination at all? Miracles are occurring for us all the time.

After receiving my driver's license, I was driving along a deserted country road which was a perfect place to practice my driving skills. It was a sunny, summer afternoon and I felt on top of the world just cruising through the countryside. I had the windows rolled down and even had my radio playing. With my left arm resting in the window frame of the door, I was trying to appear cool and experienced.

While cruising along, the engine suddenly sputtered and the car began jerking as the engine cut in and out. Just as the engine stopped completely, I was able to coast to the side of the road. Although I tried to restart the car several times, I had no success. Here I was in the middle of nowhere with no house in sight. I got out and opened the hood not really sure what I was hoping to find to be the problem.

While I recognized some of the basic engine components, I was not prepared to even attempt to look for the cause of the problem.

Standing there, on this lonely country road, and somewhat disgusted with the car and myself for coming this way, a station wagon approached. Should I try to flag him down and ask for a ride into town? I hesitated. As the car came closer, it slowed down and finally stopped behind me. It was comforting to have at least someone here who might lend assistance. A man in his early fifties got out of the car and without saying a word walked to the back of his station wagon and retrieved a small toolbox. While coming up to the front of my car, he just briefly looked at me as if to acknowledge my presence. Then, while turning towards my car engine, he just said, "Got a problem?" He had a sort of gentle smile and a quiet demeanor that immediately made me feel at ease.

While I nervously began to explain what I had encountered, he already had his head under the hood removing the screws of the distributor cap. I don't recall him saying much if anything, while I was still trying to explain why I was here. With the distributor cap removed he began working on the points by cleaning the space between the contacts. Just minutes later he replaced the cap and while straightening up, he indicated for me to give it a try. To my amazement the engine started immediately and before I could get out of the car, he had already closed the hood and was retrieving his small toolbox. Relieved and excited I began to thank him and while asking what I could pay him, he just waved his hand and said, "You're all set", got back into his car, and left.

I stood there for a short while puzzled by what I had just experienced. His car had been the only other one on this road and he happened to be the right person to fix, in just minutes, what for me would have been a big problem. I did not even ask who he was or where he lived, nor did he volunteer any information about himself. I kept wondering who this quiet and kind man was who appeared just when I needed his help. Would he ever realize what his act of kindness meant to me? Still shaking my head in wonderment, I got back into my car, grateful to be able to return home.

This experience had a memorable impact on me as an example of a simple kindness performed by a person who asked nothing in return.

How often the greatest lessons come from the simple and humble people we encounter. To me this was a true miracle performed in a moment of need rather than a lucky coincidence. I would experience many more such perfect coincidences in my life and so I am grateful for this valuable encounter which taught me this simple but valuable lesson. Even now, I still wonder about the fortunate sequence of circumstances that had brought all of this about. It would be easy to write this off as just a lucky coincidence but, for me, this would simply be ungrateful.

The Imbalances of Abundance

We already touched upon the question of whether the universe may have a long-term plan or a role to play for each of us. If we assume this to be the case, then it becomes difficult to understand why some people are successful while others fail. Why are we so fortunate to live in a country of abundance while in other parts of the world people do not even have clean drinking water, the most basic component for survival, readily available to them? With our limited human abilities, we try to find justification and reason for these enormous imbalances. I am certainly no exception in the desire to understand this extreme discrepancy. It is generally believed that the richer nations of the world have the ability to feed all of the less fortunate. I am doubtful, however, that this theory could be maintained indefinitely, if at all, especially in view of the past growth-rate of global population. According to an article in the New Scientist, "Global population forecast falls", populations are being stifled and are headed into decline.

These are questions for which we have no real solutions. Even our religious leaders, who often claim to have greater insights, find themselves struggling with this issue. It is quite possible that the simple reason for our existence is just to serve God and each other. A simple purpose, yet one in which we continue to fail. Instead of accepting this basic premise that we are all part of God's family, we bicker about who are his true followers and so continue to remain divided. How then

The Universe, is it Guiding Our Lives?

can we ever hope to solve the world's problems when we can not even accept and respect each other for what we believe?

Again, this book is not intended to answer this unsolvable question. It would be unfair, however, to imply that any predestined plan only consists of the good and memorable aspects in life. It needs to be acknowledged that each of our lives includes both joyous as well as troubled times. It is this diversity that distinguishes us as individuals. Although I have elected to point out the more positive aspects, I too have faced disappointments, setbacks and devastation. I will touch upon some of these later.

This global problem causes us to think. If the universe stands ready to help us, why then do we have such imbalances? Although we have no answer to this dilemma, we often have a tendency to view the rest of the world from our own perspective. In doing so, we fail to take into account other cultures and circumstances. When we begin to examine the situation further, we find that much of the world's upheaval is created by humanity itself. The lust for power and the desire to control, often connected with greed, have been the causes of wars, devastation and suffering throughout history. Although we may look at the universe and ask why, much of this is the result of the misuse of our God- given gift of the ability to choose. Naturally we may ask, "What have I done to cause this problem?" When, in fact, the question we should ask ourselves is, "What have I done as an individual to alleviate or improve the situation for my fellow man?" This may well be the message the universe has for each of us; namely, to respect and support each other with the gifts which we ourselves have so fortunately received.

We have a general tendency to assume that anyone who has less than we do is worse off. Much of this is derived from the fact that we often equate happiness with having possessions. We have become a society that never seems to be satisfied with what it has. Our homes need to be larger, we want better cars, bigger flat screen TVs, more up to date cell phones and to top this, we want to dine out more often. We find ourselves in a constant race for more and better. As a result of these cravings, we have become a nation of credit spenders which in turn has caused us to live in fear of losing everything. Our moments of happiness usually only last us until our next want. Can we really claim that we are happier people because of our abundance or have we become slaves of our own cravings? Do our concerns over protecting our assets cause us to be selfish and less giving?

Occasionally disasters or hard times become reminders of what is really important in life and so we can see that they too serve a purpose. These are often times when we see the real goodness of people emerge. I recall surviving the aftermath of the Buffalo blizzard of 1977, when we gathered with two other families in the basement of our neighbor's home. They had the only gas stove to cook with and to provide us with some heat. These were days when we found ourselves just trying to survive. I recall them also as memorable days as we were joined by necessity but, in surviving them, we became better friends and neighbors.

Clearly, we have a tremendous ability to learn. However, unless we are periodically reminded, we either become forgetful or simply ignorant of the world around us. Tragedies or circumstances often become reminders of our purpose and the role we are intended to fulfill. Here, too, I believe that the universe serves a purpose to encourage us to do good. With our free will, however, the ultimate choice always remains ours.

A Young Boy's Dream

Sometimes the universe takes its time in the fulfillment of our predestined plan but it never forgets. The following two examples, although perhaps of no great significance for the reader, were nonetheless reminders for me that the pieces of my destiny were still coming together.

In the early part of 1963, I received my draft notice to report for military duty in the US Army. The Vietnam Conflict was in its beginning stages and so the draft was in full force. After completing my basic training at Fort Knox, Kentucky, I was assigned to the 3rd Army unit at Fort Campbell, Kentucky, a support company to the 101st Airborne Division. I was quite astonished when I recognized the 3rd Army shoulder patch as the same one that had been worn by the American soldiers who had once occupied my home town in Germany in 1945. Now, some eighteen years later, I was part of the same army unit that I had once admired as a young boy.

In the military, each soldier is trained for a primary function known as an MOS. Mine was that of a machinist, a trade I already had acquired prior to entering the service. For whatever reason, I was given an opportunity to become involved in helping coordinate the weekly troop training session held each Saturday morning. It was my responsibility to select and obtain the various training films and other material needed for these sessions. Although it was intended as a part-time activity, it soon demanded the greater part of my duty time. As this activity required on-post travel, I was assigned my own jeep and allowed to set my own schedule. As mentioned in an earlier chapter, by watching post-war army units as a young boy in Germany, owning a jeep had been one of my early childhood fantasies. Now, through some rather unusual circumstances, this part of the greater plan also materialized.

Following my two-year tour of duty, I was more fortunate than many others by being able to return to my previous place of employment. By now my English language skills had reached a comfortable level and so I was no longer at a disadvantage. While serving in the military, I had come to recognize the clear class distinction between the officers and the enlisted level personnel. It was something that caused me to think about the level I had hoped to achieve in the workforce. Influenced by this analysis, I decided to get a college degree while continuing to hold my full-time job. Although I was faced with some difficult years ahead, just having this opportunity was far more than I had ever hoped for. Again I believe that God's plan for me was greater than my own limited vision. The long-term benefits I received as a result of this decision and the sacrifice it required were life changing and important to my new family.

Continued Blessings

Starting a family brings with it new needs and demands. I felt fortunate to have a steady job and, in fact, had been promoted to the position of section supervisor in the machine shop and was now in charge of sixteen people. We had been living in a small apartment and with the birth of our first child we were now looking to purchase a small house. After looking at approximately forty-five houses and having become somewhat disillusioned with what we had seen in our price range, we decided to have a house built in a moderate new development. As is the case with most young couples, accumulating enough money for the down payment to satisfy the mortgage requirement was, of course, a real concern. It truly took every penny we had saved, but finally a closing date was set.

Unknown to us, however, the first year's property taxes needed to be paid in advance. This was totally unexpected and we had no hope of raising the money within the required time. We feared that taking out an extra loan could possibly jeopardize our mortgage approval. Several days passed and we still had no hope of acquiring the needed amount. Now, of course, we became concerned about whether we needed to postpone the closing. We had also cancelled our apartment lease and so were in a real quandary. Just days prior to our deadline, something truly miraculous occurred. While checking our mail, I found a letter from the government. To our surprise and great relief, it contained a check for more than the necessary amount to cover the required property taxes. This was something completely unexpected but most timely to avoid our much feared crisis. Receiving this tuition reimbursement check was nothing short of miraculous as the timing and the needed amount could not have been more perfect. I could not help but wonder whether the universe had been intervening again or whether this had just been another lucky coincidence.

These special graces or gifts come to us in many forms. It is most surprising when they show themselves in the right form and at the most appropriate time. Occasionally even those situations which we may consider disappointments often turn out to be a blessing in the long term. Our inability to see beyond the here and now often causes us to miss opportunities. This is probably best expressed by the slogan, "it's now or never." I frequently find myself in this short sighted category when, instead, I should realize that there may be a reason for a more appropriate time. I am certain that most of us have tried to go in directions for which we were simply not suited. I recall when I applied for jobs either to earn more money or just to do something different; although those were not bad motives, they did not feel right. Naturally I was usually disappointed when the job I had hoped to get was given to someone else. It was only in retrospect that I could see that my previous attempts to improve myself would have derailed the course I was meant to take.

I was still attending the university and this also had raised my hopes of improving my position within my company. Ultimately, an opportunity presented itself that would take me in a new direction. I was awarded a position within the industrial engineering department and so, had an opportunity to work closely with middle management in an office environment. During the same time, however, the firm was also experiencing a severe economic downturn and with it came the threat of layoffs.

Just days after I had started my new job, employment cuts were being implemented; the entire industrial engineering staff with the exception of myself was being reassigned to other departments. Not only did I survive another layoff, I was now given the task to carry on the basic functions of the department. Although it appeared to be another just-in-time miracle, I would eventually realize that it had set me on a course that would offer further opportunities for advancement in the years to come.

People frequently identify with their work, and so losing their jobs can not only result in losing their dignity, they frequently feel stripped of their self worth as providers. This combination quite often leads to depression and related symptoms.

I considered myself fortunate that in spite of various lay-offs as the result of economic downturns, I was able to continue working. In fact, on several occasions when I was worried about being laid off as well, I received small advancements instead. Each time this occurred, I just wondered with amazement. To work for over forty years without a single layoff was truly a miracle and a blessing. Why was I so lucky and who was there always looking out for my well being?

The company had already established several facilities in both Europe and Japan and was now looking to expand its global sales. By then I had graduated from the university and with my German language skills, joined a newly formed international market development department, a three man team that reported directly to the VP of international operations. It became our task to develop overseas markets, establish distributions, and ultimately support the establishment of new facilities in countries where it seemed appropriate. Because of the potential overseas travel associated with this department, my position became the envy of many coworkers.

It is interesting to recall that my first overseas business trip would take me to my former home in Germany. Again my first reaction was, what a lucky coincidence that of all the places in the world, I was able to visit my former home. While I attempted to consider the probabilities of all this coming about I felt simply amazed by the succession of events. Eventually I would have the opportunity to visit many other countries; my life-long dream of traveling the world would finally become a reality.

In addition to meeting many interesting people of differing cultures, I gained many friendships in the process. Standing on Mount Olivet overlooking the city of Jerusalem, visiting the Taj Mahal in India, standing next to Corcovado, the statue of Christ overlooking the city of Rio de Janerio, and numerous other famous sights during my travels was truly the fulfillment of a young boy's dream to see the world.

Fred Bonisch

With the increase in responsibility also came the financial benefits which would later enable me to provide the means for my children to attend the college of their choosing.

Besides the ability to travel, I now had the opportunity to meet and interact with people globally, including our own management personnel. What was unique was the unusual process I had gone through to arrive at this point. For the first time in my life, I really enjoyed all the things I was now doing, even though my responsibilities frequently occupied me well into the night.

I recall my disappointments during previous years, when other jobs I had applied for were given to other candidates. It was only in retrospect that I realized that my direction must have been predestined all along. Not only did I enjoy my activities, but I felt that it suited my personality. Perhaps because of my European background, I felt comfortable meeting with people from other parts of the world. Over time, my activities became so diverse that it was often difficult for me to even describe what it was exactly that I was doing, but I enjoyed it nonetheless. When my youngest daughter was asked during show- and-tell at her school about what it was that her dad did, for lack of a better description she told the class "My dad goes out to lunch and dinner a lot."

Much of this was to change. As the result of a change in upper management, the department, of which I was a part, was gradually dissolved. This placed my future in real doubt. Again, instead of being demoted or let go, the department was renamed and I was assigned to manage its operation. I escaped severance, and again asked "Who out there has been protecting me all these years?" Since it was also a time when my children began their college education, having steady employment and the income to support them was even more important.

In retrospect, it seems almost unreal to think of the route I was directed to take which often appeared with little hope for a future. Out of each step, new possibilities developed that helped me along my career. Could I have possibly been alone in this unusual journey? Was I just a lucky individual or was the universe in fact guiding my life throughout all of these years? It seems like a great deal of luck to have avoided the many layoffs; (not to mention the advancements) I experienced. Instead of regressing these events allowed me to move towards something that would give me greater satisfaction while at the same time continuing to provide for my family.

Our Tragedies and Blessings

We don't know why things happen nor are we privileged to know or understand God's plan for us. This frequently becomes the cause for doubt, especially when we experience difficult situations. All of us experience them in one form or another. Books have been written about "Why do bad things happen to good people?" When we consider the disastrous impact of the tsunami, even as believers we can't help but ask, "Where was our loving God in all of this?" Or, "Where are those good spirits when we need them most?" During those times, we tend to look to our spiritual leaders for answers only to find that they too are at a loss for answers. If what occurs is part of God's plan, then our spirits too must conduct themselves in accordance with His will.

With our human limitations, we can only see our past up to the current moment and so cannot see what lies ahead. Often our tragedies make us stronger or bring us together. We frequently wonder why some die so young while others live to old age. Those whose faith preaches re-incarnation believe that we actually select our kind of life before we are born. By doing so, the believers hope to achieve higher levels of perfection by selecting a more difficult life for themselves each time they return to earth to begin their new lives. These re-incarnations continue until finally, after having reached the highest level of perfection, the believers no longer need to return. Although this theory would explain the reason for much of our suffering, it is, however, contrary to the beliefs of most of the world's major religions. This then brings us back to the realization that we simply are not meant to know our future nor the reasons for our longevity.

As mentioned in an earlier chapter, believers find their hope in the promise of an afterlife. We view this as a spiritual life of perfection while being in the presence of God. We are taught that all of our earthly problems and needs will no longer burden us as we too become part of the spirit world or the community of saints. It may be of interest to point out that we are not intended to become angels as God created angels separately and for his own purpose. Although we have the promise of salvation and of living with God forever in what we so freely refer to as heaven, rarely ever do we hear of anyone rushing to get there as a result of this great promise.

Rick Warren, in his book *The Purpose Driven Life*, points out that one of our purposes for being on Earth is for procreation and to establish family life. This confirms our need for community, with the family unit being the one most fundamental and essential component of our communal life. It also fulfills God's basic instructions to mankind when He advised Adam and Eve to "be fertile and multiply." (Genesis 1:28) When young people first start their own families, they quickly learn to no longer think only of themselves but also about the well-being of their families. Through this growth process, which we recognize as unselfish love, young parents place other family members ahead of themselves. The family unit then is crucial since it becomes the base of our earthly foundation. Those who lack this most basic unity during their early childhood years often display unusual behavior in later life.

As a family we are no longer a single entity. We share in the joy and the disappoint- ments of each other. This means that we not only deal with our own issues but those of others as well. This, of course, can multiply our joys but can also worsen our painful experiences. What one member feels affects all the others to some degree. We call these feelings love as well. Although creating a family may be our most important purpose and a part of God's plan for us, it may not be all we are destined to fulfill or to achieve.

For me as a child, my family was always a place of safety and comfort and a peaceful haven even during the dark years of World War II. Although I have always had a strong desire to travel, home was where I felt a real peace. It is perhaps the only place where we can truly be ourselves without pretence. Ultimately we take on multiple roles such as child and/or parent as we begin to create our own family. Much of the comfort we experience is carried into our new family. However, as parents we now take on new roles as protector, provider, and teacher. We usually begin this important responsibility without much preparation or expertise other than to rely on our own childhood family experiences.

I was married at the mature age of twenty-nine while still holding my job and attending college. Within a year and a half we were blessed with our first child, a baby girl we named Kimberly. It would be four years later when we were blessed again with a baby boy whom we named Marc Christopher. With this happy addition, we required more living space and so purchased our first house. I had finished college and was advancing in my job, so we considered ourselves quite fortunate.

Several months before Marc reached the age of two, things changed very drastically for our family. Marc would wake up crying almost every night, but he was still too young to tell us what was bothering him. He became pale and less energetic. After several doctor visits and a checkup at our local children's hospital, Marc was diagnosed with leukemia. Our peaceful world had just crashed around us as fear of losing our little boy totally consumed us. Doctors who treated Marc gave him only a twenty percent chance to survive. These were not very positive odds but we clung to them with all our hope.

Needless to say, I searched my life wondering what I had done wrong to deserve this pain. We asked God why; although there was no answer from the universe, we received much support from friends and family.

The local children's hospital immediately began Marc's treatment. He needed to remain there for the next several days. With the start of the treatment he became so weak that he could barely sit up. Every fearful thought was followed by a short silent prayer. Our friends met in nightly prayer sessions dedicated to Marc's recovery. After his release from the local hospital, he was scheduled for regular treatments at our local cancer center. Each treatment usually required an overnight stay on the special children's floor. Soon we found ourselves becoming acquainted with most of the other parents and their sick children.

Seeing those young boys and girls afflicted with various types of cancer was truly a test of faith, yet what else was left to support our hope. It seemed that each time we came back for his treatment, another child we had come to know had died, and each of those times I was stricken with panic and a tremendous feeling of fear.

In trying to muster every possible resource, we asked our local priest, who also performed healing masses, to pray over our little boy. Marc was again receiving a treatment at the hospital and was simply exhausted from the medication. With Marc's limp body lying in a special movable bed cart for children, our priest began to pray for our son. Shortly thereafter Marc raised himself to a sitting position. Although a small sign, this event remained with us as unforgettable.

Marc remained in treatment for several more years as the leukemia cells continued to appear in his painful bone marrow tests. Other children from the same hospital ward whom we had come to know so well, continued to die and our hearts sank with each passing. As we sought every ounce of hope, each such passing was like being hit in the stomach. Although I prayed as never before, I realized that all of the other parents we had come to know were doing the same and yet their children continued to die one by one. How could I dare to hope? Why would God make an exception for us?

Just prior to Marc's fourth birthday, we had another addition to our family, a baby girl we named Megan. We had also purchased a larger home so that each of our three children could have their own room. Since they were all still quite young, the doors to their bedrooms were kept open at night so that the hallway light would illuminate each of their rooms. The open doors would allow us to hear them in case of a problem and also permit them to find their way to the bathroom during the night.

One morning, Marc came into our bedroom and while leaning over the edge of the bed, he told us about an experience he had during the night. He spoke of a man that had come into his room and stood at his bed. He described the man as wearing a long, white gown. My wife and I looked at each other incredulously. Marc stopped speaking as if this was the end of his experience. So I asked him, "Well, what happened then?" Marc continued to tell us in a very peaceful manner that the man then picked him up and held him in his arms. "Shortly thereafter he laid me back down and went away." I waited a few moments to absorb what we had just heard. I remember asking Marc whether he had not just dreamed this whole thing. He became angry and assured us that it was not a dream and that the man had really come into his room and held him. My wife and I looked at each other without either of us saying a word.

I couldn't help but think that if Marc really had this experience, why wasn't he frightened by it? There was enough light in his room to allow him to clearly see. Suddenly, and I was not certain why, I believed him, but wondered who it was that had come to visit him during the night. Could it have been Jesus or an angel or some other spiritual being? We would never know and yet we realized that something extraordinary had happen to our ailing boy. Although Marc never spoke of the event again, the experience will remain with me forever.

Marc eventually recovered from his illness; meanwhile, many of the other sick children had died. I often wondered why our son was spared but am grateful to God for having done so. In 2008, Marc was a thirty-three years-old spirited young man. He is married to a wonderful woman and both were blessed with their first child, a beautiful baby girl they named Katelyn. In spite of what he had to overcome, Marc was able to follow his dream to graduate from an aeronautical university and to become a commercial pilot working for a major U.S. airline. Here the reader may recall my earlier expressed desire to fly which had led me to my private pilot license. I like to think that my ambition to fly set the stage which would ultimately be fulfilled when Marc became a commercial pilot.

Perhaps it was Marc's illness that influenced our oldest daughter, Kim, to become a medical doctor; (a pediatrician). Our youngest daughter, Megan, became a teacher and guidance counselor. I am proud of all my children and feel very blessed, not just for their accomplishments, but also, for the caring and responsible people they have become. It is interesting that three children from the same parents should take such different directions. It simply reaffirms that God truly has a unique plan for each of us and, if it is His will, His plan will be achieved regardless of the difficulties encountered along the way.

It is with pride that I realize that each of my children serves humanity in a different way; namely in the fields of healing, transportation, and education. For God to fulfill this portion of His plan, so that these particular children would be born, I was destined to come to the United States. Out of a population of approximately two hundred twenty million people, I needed to meet that one specific person in order to bring forth these three unique individuals.

The readers may note that for all of this to occur it required proper sequencing and timing. Did it simply all happen by luck or coincidence or did the universe, in fact, guide the events? We can't say with certainty but it should cause us to think and wonder. The extraordinary occurrences continued as I will describe in subsequent chapters.

Seeking Intercessions

Miracles and blessings can come to us in many different forms and circumstances, which perhaps is the real blessing. People often expect their blessings, help, or miracles to come in a certain way or through a particular source. Calling on St. Anthony when something is lost, for example, has become a standard practice for many Catholics. This, however, is not viewed as an acceptable practice by most other denominations.

I recall an occasion when I sought the help of St. Anthony when we tried to terminate our safe deposit box. We had just returned from the local bank where we had emptied our box. After we arrived home we realized that the key to the box was missing. Although we had thirty days in which to find it, we began to look for it immediately. Not finding the key would result in the bank having to replace the lock at our expense. After having looked unsuccessfully for about three weeks, the situation became somewhat desperate. Since we had already checked every possible area twice, I decided to pray to St. Anthony to help us find this missing key. Almost instantly something told me to search the right side of our rear car seat. This was a specific reply. I had already checked the car several times before without success. Upon receiving this clear sign, I went directly to the car and slid my hand between the back seat and the side wall. Almost immediately, I felt the metal portion of the key and extracted it. The response of my request to St. Anthony was clear and prompt. Strangely, receiving the prompt

Fred Bonisch

answer to my request did not come as a surprise at all. He had helped me several times in the past. Having had these positive experiences with my favorite spirit saint has made my relationship with him a most natural and trusting one.

Having received such positive and personal responses by calling on St. Anthony, I do not find it unreasonable to think that there are others in the spirit world who might respond to our needs as well. As Catholics, we already have saints for nearly every situation. Several years ago someone gave me a medal of Our Lady of Loretto. A message on it reads: "Patroness of Aviators & Air Travelers, protect my flight." Since the Church has declared particular spirits saints, we call on their assistance and intercession for our specific needs. Here again the practice points clearly to the belief by Catholics in the spiritual ability to assist us. Since we have discussed the saints in an earlier chapter, we need to mention nothing more about them.

Although I am not certain, Catholics may constitute the largest religious group that believes in a place known as purgatory, assumed to be a place where the souls spend time after death before they are allowed into the presence of God. The Church believes this to be a period of suffering and cleansing. During past centuries the Church described this suffering as similar to that of Hell which could last for months or even years, depending on the individual's previous lifestyle. The perception of this period was described as so painful that many believers sought forgiveness through the Church by means of indulgences. This ultimately led to the Reformation and the split of the Church in Europe.

In more recent years, the church has aimed to change this old fearful perception by redefining what is now meant by the suffering in purgatory. Cardinal Joseph Ratzinger, former head of the office known as the Congregation for the Doctrine of Faith (CDF) at the Vatican prior to becoming the current Pope, redefined purgatory in his book on (1) *Eschatology, How to Explain Purgatory to Protestants*, as: "…. *the inwardly necessary process of transformation in which a person becomes capable of Christ, capable of God [I.e.' capable of full unity with Christ and God] and thus capable of unity with the whole communion of saints……….*" P-13

Although the perception of purgatory still varies among the laity and the clergy, it defines a process by which the departed ultimately become united with God. We see the journey by which a person, upon death and through the process of transformation, ultimately joins the communion of saints. This, however, does not necessarily confirm that they too will become saints. It seems to at least indicate an equal status with them. If we accept that we can call upon the saints to intercede for us with God, then it would appear reasonable to think that we can also call on any of our dearly departed to possibly do the same. Again, this concept will be rejected by many in the Christian world as simply unbiblical. The Catholic Church believes that not everything is stated in the Bible and that the Holy Spirit still continues to guide the Church. For those who do accept this concept, it would substantiate the belief in the existence of a spirit world, a universe that may stand ready to assist us.

Why Do Things Happen

Much of what was described in previous chapters refers either to myself or to my family. This is not meant to imply that I walk under a golden cloud or that I am someone special. I have not been given special abilities or powers. I am thankful to whomever it is that is watching over me and for being there when the need arises. With all the blessings I have received, I still have my moments of doubt. None of us escape life without periods of pain or hurt. Losing a loved one, for example, is always a painful experience and yet it is part of life. During these periods we often feel alone and wonder whether God is really hearing us. We are beings with feelings and emotions and so we naturally respond to the events occurring in our lives.

For me to imply that everything in my life has always been positive would be misleading. I have had my share of setbacks. Sometimes they are small while at other times they have really shaken me. Like most of us, I have dealt with the death of loved ones and experienced illnesses and misfortunes. Surviving a divorce and being separated from my family was truly one of those dark periods. Often these periods appear endless and we tend to feel alone in our grief. It is hard to accept during the time of our grieving or anger that we are being helped. Feelings and emotions need to be acknowledged and allowed to be felt before we can begin to heal. As human beings we have the ability to overcome adversities and move on with life.

I once attended a Sunday service at a Bible church outside of Washington D.C. By coincidence the pastor spoke about the very subject

The Universe, is it Guiding Our Lives?

of suffering. In an effort to help us understand why tragedies happen in our lives, he pointed out a number of factors. I was particularly touched by three of his explanations and so they remain in my memory. He first explained that all of us suffer at one time or an other, and that it is simply part of life which none of us can escape. Second, and perhaps the most meaningful, is that "we are never closer to God than we are during the time of our suffering." I think that men in fox-holes can certainly attest to this. His third point which remains with me is that, "in our sufferings, regardless of whether it is mental or physical, we can become an inspiration to others." While these views do not explain reasons for our sufferings, as believers we can take at least some comfort in these explanations. While listening to his inspiring sermon, I couldn't help but wonder about how non-believers or atheists handle the tragic situations or circumstances in their lives?

We hear it frequently expressed that "life isn't fair." Perhaps there is some truth in this saying as we hear it from both the believers and non-believers alike. Just as I wrote this, five hundred people were killed by explosives in Iraq. At the same time an earthquake hit Lima, Peru, with the number of people killed expected to be in the thousands and the total damage estimated to be excessive. When we look at the disastrous slaughter of innocent human beings, often done out of hatred and total disregard for life, we find that it is frequently done in the name of the same God we all worship. If, in fact, we believe that we are all created by the same God, how then can we commit such heinous acts against our fellow human beings without insulting the Creator Himself? When we begin to view much of the world's problems, we come to realize that they are frequently the result of hate, greed, and the lust for power. These then are the satanic powers at work, a spiritual force from which we wish to separate ourselves.

If, on the other hand, we look at natural disasters such as earthquakes and tsunamis, is there anyone to blame for such events? If the spiritual force in the universe is supposed to help and guide us, why did they not intervene? We need to acknowledge that we have no real explanation for this and are not entitled to one. We could assume that we receive the help of the spiritual forces for our occasional individual needs but that natural catastrophic events remain the secret of the universe. We will most likely never know why this is so and as a result it will remain a great mystery for every generation, past and future.

The Feeding of the Hundreds

Natural disasters can strike anywhere. I live near the city of Buffalo, New York, which has a reputation of being snowy and cold during the winter months. Although we experience an occasional snow storm, most who live in this area accept these storms as normal. The great blizzard of 1977 validated much of the reputation. The magnitude of this storm has clearly remained a memory for all who lived through this most unusual event. Even now one still sees T-shirts marked, "I have survived the blizzard of 77" worn proudly. Survival is the most important blessing in the end, but we have become so attached to our possessions that their loss or damage often overshadows what should really be important to us.

The day of the blizzard started out as every other day. The morning was clear but by noon it began snowing heavily and it didn't take long for it to accumulate. As the wind increased, it turned the snow into blizzard-like conditions making driving extremely dangerous if not impossible. By mid afternoon it became clear that neither I nor any of my colleagues would be able to drive home very soon. Several feet of snow had accumulated and, with the addition of the wind, it was nearly impossible to see beyond just a few feet. Snow plows were attempting to keep the main road in front of our plant open but without success since the snow was falling faster than it could be removed.

When several hundred workers realized that they would not be able to leave their buildings, they "raided" the food vending machines.

The Universe, is it Guiding Our Lives?

Within just a short time all of the machines were completely emptied. In the meantime, the storm continued its ferocity and most of the personnel were resigned to the fact that they would have to spend the night at the factory or office. There were several hundred people with little food, since the snack- style vending machines were not intended to serve this unusually high demand.

My office was located in the front building of the plant and so I was able to look onto the main road. The drifting and blowing snow made it nearly impossible to recognize anything beyond thirty feet. While peering out of the main door window, I noticed an oversized van trying to make its way on the main road about thirty feet from our building. It had come to a stop near the main door of our headquarters facility. A large snow drift had mounted up in front of the vehicle and before the driver realized it, another drift caused by the strong wind had quickly built up behind the van. Blocked by the snow drifts, the driver found himself stuck on the main road directly across from the main door of our administrative building. Unable to go any farther, the driver got out of his vehicle and made his way through the blowing snow to find shelter at our main office building.

After joining the employees inside, he explained that he was returning from the city and was heading home to the nearby town of East Aurora. He had just picked up a van load of submarine rolls, fresh ham, cold cuts, tomatoes and mayonnaise and was heading back home to his submarine restaurant in town. Here was a man with a van filled with food, stuck exactly outside our main entrance, while inside our buildings were hundreds of hungry people unable to go anywhere. Recognizing the need and the opportunity, the company's management immediately decided to purchase the man's entire supply of food. Several employees quickly formed a human assembly line to bring the much needed goods inside. Under the supervision of the restaurant owner and several employees, submarine sandwiches were prepared and distributed and eventually everybody was fed for the evening. It was simply astonishing to realize that this man had gotten stuck exactly outside our main door, with the right quantity of food to feed everyone stranded inside.

I am certain that most people counted themselves as simply lucky that this man should become stranded at this opportune time. For me, it was a miracle in the truest sense and it reminded me of the parable of Jesus and his disciples feeding the five thousand with just five loaves of bread and two fish (Mark 6: 34-44). The probability of all these events occurring at the right time and in the proper sequence is simply beyond comprehension. This miraculous event is one of the great miracles I was fortunate to witness and to share. There is no doubt in my mind that a higher power was at work to bring this about. Perhaps because I was able to recognize it for what it was, it was meant for me to witness this great and wonderful blessing. Although the blizzard would remain an unforgettable event, no one involved talked later about this miraculous experience of the 'just-in-time sandwiches'. It made me realize how much we simply take for granted and how ungrateful we can be.

The fact that most of us have sufficient food does not come about by chance. It requires the involvement of many people (the farmer, the processor, and the trucker), until it reaches the stores and becomes available to us. In order for crops to grow sun, rain, and wind are required elements over which we humans have absolutely no control. This should cause us to think and to feel grateful when we enjoy our meals. More importantly, we need to recognize the higher power that is at work to bring all of this about; not just once but day after day and year after year.

A Timely Intervention

Although I try to be observant of these graces, I am certain that many of them escape me and go unnoticed. We simply take too many things in life for granted (good health, good friends, family). These are just a few of the blessings we usually accept as normal. Every so often I am reminded of the extraordinary events that have occurred in my life and feel assured that my guardian is still with me. As was mentioned in a previous chapter, the late Pope John Paul II credited Mary for his survival of the gunshot inflicted by his assailant. We each may have our own thoughts about whether his claim is correct or whether it is just an assumption on his part. The fact remains that he, being a person of the highest religious position and well respected by many throughout the world, believed very strongly and openly that the intervention by a spiritual being played a role.

The following event has actually little to do with religion and is of no great practical consequence, yet for me it was something so extraordinary that I feel it worth mentioning. It is an example of the diversity of assistance we can and should expect from the universe. My youngest daughter Megan, then about age seven, was playing with a friend inside our house. Her friend's father, whom I had never met before, was scheduled to fetch Megan's playmate. While the girls were playing in the family room, I walked into the basement wondering what to do about my defective sump pump. It had started smoking a few hours earlier. I had disconnected and removed it from the sump hole and laid its parts out on a nearby table. I had never before had to deal with a sump pump and was, therefore, somewhat at a loss about what to do with it. I began to accept the fact that I would likely have to purchase a new unit. Since it had been raining intermittently, the sump hole would surely fill up quickly and possibly flood the basement. I realized that it was not something that I could delay for more than a day without suffering the consequences of water damage.

As I mused about my situation the doorbell rang and, as expected, the young girl's father arrived to take her home. While the girl gathered her things, we introduced ourselves. We began to talk and my misfortune with the sump pump entered into the conversation. He showed an immediate interest and asked about the pump's make. He told me that he worked for a local company which was the only company in the area that repaired sump pumps. To my great surprise, he mentioned that he was, in fact, the person who performed all the repairs on this make of pump.

I led him into the basement to show him the failed unit. His response was simply, "Oh yeah, we can fix this. Just bring it in first thing in the morning and I'll take care of it right away." Although I didn't ask, I wondered whether he was the sump pump angel who knew just when to appear on peoples' doorsteps.

He did take care of the broken pump the next morning and I had it back in the sump hole before noon. This timely coincidence still puzzles me and I have told this story to many of my friends. I had never met anyone having anything to do with sump pumps, yet at the very moment I needed help, the only pump repair person in the area appeared at my door. I have not met a pump repair person since, which makes this once in a life time experience so extraordinary.

Certainly the event was not life threatening but the circumstances under which it occurred, gives one of the clearest demonstrations of an invisible outside force. Whenever a discussion of our unexplainable events arises, this is one of the examples that comes to mind. I must admit that I am still in awe when I recall this perfectly-timed coincidence.

Trusting With Reservations

If we believe that there are spiritual beings on whom we can depend in the universe, regardless of whether we call them saints, spirits, or souls, we may find ourselves returning to our earlier question of why some of our prayers are answered while others are not. Why do certain things happen even without praying or asking for them? Is prayer then really helpful or necessary? Jesus assured us (John 16:23) when he said: "...*whatever you ask the Father in my Name He will give you.*" He also advised His disciples, and us as well, that in order to receive assistance we need to believe. I see this as a hurdle even for strong believers.

Certainly there are individuals who will find their own reasons for unanswered prayers. Although we may have the best of intentions, doubt usually enters our shallow trust.

When we address our unseen God, whom we believe has control over the entire universe, we do not normally receive a confirmation or a response and so we begin to doubt. I have encountered clergy who struggle with this issue just like other believers. It is in our human nature to always approach things with a certain degree of caution and skepticism.

As evidence of our lack of trust in our prayers, the following question was asked of the seven Democratic candidates running for the presidency of the United States during one of the open panel debates in 2008: "Do you think that prayer could have prevented the 2004 tsunami or the 2005 Katrina disaster?" In their responses, each of the candidates first acknowledged that prayer was part of their lives. I found it interesting that by acknowledging their prayer life, they simultaneously acknowledged their belief in God and did so openly. It is important to realize that these people are not members of the clergy but rather political candidates running for one of the most powerful offices in the world.

It is understandable that prayer life varies greatly among people but the candidate's unanimity on this issue was somewhat astonishing. What was more surprising was that each of the candidates acknowledged that they thought prayer would not have prevented either the tsunami or the Katrina disaster from occurring. A few of the candidates expressed that they felt that prayer and faith helped in the aftermath by bringing people together to assist those in distress. I do not wish to judge whether their expressed views were a political ploy or a response from the heart.

It should be acknowledged that these candidates do not represent the view of society. It gives us a hint, however, of their perspective, considering the fact that these men and women are part of the leadership of our country. What is unique, especially coming from a diverse group of politicians, was their unanimity of opinion that prayer would not have prevented these extraordinary disasters from occurring. In retrospect we may wonder whether they would have prayed had they realized in advance the enormity and the destruction which resulted from these occurrences. Unfortunately, that question was never asked.

While we can criticize the candidate's responses, we each have situations when we expect prayers to be answered and others when we do not. I doubt whether our general thinking is really so different from theirs. Having complete faith in prayers is probably the exception rather than the norm. To be doubtful of promises is one of our human traits of safeguarding ourselves against disappointments. We are even affected by this trait when it involves a promise from God Himself.

The tsunami was an unexpected disaster of such enormous proportions that there was almost no opportunity for advance prayer. With Katrina on the other hand, we knew it was coming but the outcome was more severe than was feared. Would prayer have prevented this disaster? I doubt whether even our religious leaders would be able to give us a clear yes or no answer to this question. When it became evident that the hurricane was approaching the New Orleans coast, I am certain that many of its residents prayed for this storm to prove less damaging than anticipated. Doubting, even to a small degree, is not what our religion teaches us and yet it is a very natural human tendency. Although a few religious leaders have suggested that the hardship imposed, at least by the Katrina disaster, was a punishment from God, it is certainly not a position shared by this author. Since no one is privileged to know God's plan, we need to admit that we simply should not make such judgments.

Let's place ourselves in God's position for a moment just to see how this question looks from His perspective. If He were to answer all of our prayers and requests, we would, in fact, be able to control our own destiny. We would simply ask and it would be done. Some would unquestionably try to control the destiny of others or even that of whole nations. Whether we like to admit it or not, we are to some degree selfish, greedy, envious, controlling, and lusting for either power, or recognition. If everything were granted to us, we would soon try to turn God into our personal genie. By looking then from this perspective, we can see that God needs to be selective in what He grants or denies us. Without knowing our future, we have to admit that we cannot know what is best for us. What may sound wonderful at the moment may be detrimental for us in the future. If we are helped by the spirits, they will know God's will and so will treat us accordingly.

My own response to whether prayer would have prevented these disasters would be the same as those given by the presidential candidates who answered somewhat under pressure. In fact, I quite often find myself in doubt about whether my prayers will be answered and so it becomes a sort of waiting game. At times, perhaps out of desperation, I have even resorted to the offering of donations or the reciting of the rosary in order to achieve the granting of my prayer and so have tried to barter with God. Imagine me trying to barter with the ruler of the universe. In doing so I can almost see Him smile while saying, "Oh, you little man, I have given you all that you need. I have always provided for you. Why are you trying to bribe Me with your petty offerings?"

Why then do we pray if we have such doubts? Prayer is our way of communicating with God and we do it for many reasons. First and foremost it should be to praise Him as our creator. "The Our Father," as taught to us by Jesus himself, is by far the most perfect example of the reasons for prayer and encompasses all of our major concerns in a most perfect order. Prayer keeps us connected to Him and His universe but it also gives us hope while here on Earth for a life to come. I wish to reiterate what was said earlier, namely believers, regardless of their religion or faith, have hopes that non-believers don't perceive in the same way. We all live in the present, but for believers there exists the hope for a glorious future. We need to frequently remind ourselves of that fact.

God has always provided all my needs. It is my wants that usually get me into trouble. Many times when I was in need, something good happened just in time. It is easy to forget those events because our eyes and minds are usually concentrated on our next needs and wants. This is one of our human failings. Realizing that He has created us, God fully understands.

Difficulties We Face

As a young man from Germany who had arrived with just a basic education and little knowledge of the English language, I was pleased with what I had achieved. I was able to receive a college degree and held a respectable position that allowed me to travel around the globe and thus fulfill one of my greatest dreams. Although I was not wealthy, I felt blessed with a beautiful family and a comfortable home. I had been able to do so much and had the opportunity to meet so many people from around the world. I felt that I was enjoying the American dream (at least from my perspective). Could all of this have come about because of my own initiative and by simply being in the right place at the right time? It is easy to arrive at such conclusions as progress usually occurs over a period of time rather than all at once. However, I feel confident that I had a great deal of invisible help along this unique journey. Because of our inability to understand these miraculous interventions, I have questions that remain unanswered to this day.

As often in life, good things can come to an end. I too was going to experience one of the downsides of life. A divorce was not only a shock to my family values but to my religious beliefs as well. Suddenly much of what I had worked so hard for had changed. Fearing how the divorce would affect my future, as well as that of the children, was a new concern that I now had to deal with. Repeatedly I asked that all familiar question, "Could this possibly be part of God's plan for me, and if so, why?"

Fred Bonisch

Our married friends who did not want to take sides, understandably, withdrew. Even my own church ignored my dilemma in spite of my many years of service and involvement. As I began to withdraw from my religious activities, I felt out of place in my own church. My faith in God was all I had left; I even began to have doubts about that. In time, however, I came to realize that I was probably oversensitive during this difficult period of transition and what really mattered was my relationship with God.

My criticism and misgivings may well be perceived as sour grapes. After all, the church did not cause my divorce. However, the very institution upon which we rely during such troubled times, did little or nothing to help me survive the process or achieve recovery. In fact, since divorce is looked upon as breaking one's sacraments, it is often mentioned in context with the other downfalls of the world such as drugs, alcoholism, and many other abuses. This certainly did not define my immediate future as being part of something glamorous or uplifting. Since I was still the same person as before, I began to wonder about how much we as a people really matter to the church. As humans we expect to be disappointed by others, the church, political units, friends, families. As individuals, we too inflict pain and disappointment on others, often even those unknown to us. As earthlings we are all prone to fail. I trust that it is during these low periods in life that we should count on the spiritual strength to help us regain our balance and self respect. I can attest that it has been of great value to me.

Feelings and realizations, no matter how hurtful they may be at times, can also be a guide and even a gift. The experiences, both the positive and the less so, have enabled me to look at the world around me with more clarity. Things are not always what they appear to be. This is something we all experience at one time or another. We also have our weaknesses and strengths. It helps to know that we can call on God and His universe to stand with us at all times. In spite of the life-changing divorce, I have survived, (hopefully a stronger person) and have moved on with my life. By the grace of God, my children have turned out the best I could have ever expected, and so I feel blessed. We have all learned from the difficult years.

I am reminded of the famous picture of two sets of footprints in the sand that suddenly turned into just one with Jesus saying: "This is when I carried you." I believe that we have that special support during those difficult periods in our lives. While there may be no quick fixes, I think that God and our good spirits stay and feel with us and assist us on the road to our recovery. Only when we lose or give up on this trust, do we become lost and alone.

It is my hope that through sharing this period in my life, I have shown that in spite of the many blessings I have received, I am by no means the golden boy. I am just another humble human being trying to find his way and the role I am destined to fulfill.

Could Jesus be Chinese

The variety in which the unusual occurrences have manifested themselves has always been of great mystery to me. That they have followed me in my travels has been of great confirmation and comfort. Some of the following events were truly life threatening and yet I was spared during each of those occasions. Although there are many others of lesser severity, the few described here will make the point.

On a flight from Washington, D.C to Rochester, N.Y., our aircraft developed an engine fire at about mid-point in the trip. The fire was extinguished and the aircraft was able to complete its flight with just one engine operating. Needless to say, the passengers became extremely anxious. This anxiety intensified as people observed the flashing fire-fighting equipment on the ground awaiting our arrival. We landed safely and feeling relieved; everyone on board cheered and congratulated the pilots for a job well done.

On a flight to South America, the brakes on the airplane seized on take off. This would make the landing in Rio De Janeiro, Brazil a rather dangerous one. The aircraft was able to land but came to a screeching and smoking halt in the middle of the runway. With the brakes completely locked, the craft was unable to taxi any farther. The passengers had to exit the plane at that point, but, thanks to God, no one on the aircraft was injured.

The Universe, is it Guiding Our Lives?

In other incidents, I survived both a car crash and riots in India. I firmly believe that my good spirits were there to protect me during those fearful occasions.

While I have had several other such close calls, the following was one of my more memorable travel experiences. On a return flight from Japan, I stopped in Hong Kong to meet with the manager of our newly established sales office. On the morning of my last day in this fascinating city, I awoke with a severe lower back pain which made it difficult to move or sit. This was something I normally experienced about twice a year and it would usually take about a week before I returned to some normalcy. I was greatly concerned about handling my heavy luggage and surviving the sixteen hour flight back home the following day. After an early business meeting at the hotel, I tried to rest and soak in the bath tub in hope of getting at least some relief. By noon my body took on the shape of a question mark as the muscles tried to compensate and relieve the pressure on the nerve in the vertebrae. A dinner meeting had been scheduled for the early evening and so I had most of the day to rest. By early afternoon I decided to venture out of the hotel hoping to find some gifts to take home to my children. My back problem had worsened and I was now walking with my back totally out of alignment. I was concerned about venturing too far from my hotel, which fortunately was directly in the heart of Kowloon, the most common shopping district of Hong Kong.

As I made my way along the busy downtown sidewalk, I noticed a beggar standing only a few feet away from me. He was a middle-aged Chinese man leaning against a store front and supporting himself with crutches. Although I have encountered many beggars in my travels, this man caught my attention. I stopped and was nearly run over by other pedestrians. I patted my pockets for change and pulled out all my available Chinese coins. I turned and took a few steps towards the man. As I approached him, I noticed that part of his left foot was missing. His clothing was dirty and his crutches appeared well worn. I dropped the few coins into the metal cup he held out. As I did so, I wanted to acknowledge him as a human being and so I looked straight into his eyes as the coins dropped to the bottom of the cup.

His face looked older than I had first thought. His skin was wrinkled and brown and showed signs of a hard life. His face was unshaven, giving him a somewhat unclean appearance. During the brief moment that I looked into his face, focused on his eyes, he acknowledged me and said something in Chinese which I did not understand. I assumed it to be his way of thanking me. All of this occurred in just a few seconds. As I turned to walk away, the man's face was still on my mind and it wasn't until several seconds later that I realized that my back pain had vanished. I was walking fully upright and there was absolutely no discomfort. Astonished by this sudden healing I wondered what had just happened that caused this sudden recovery.

It was not until I had returned to my hotel room that I realized the full significance of the enormous miracle I had experienced just an hour earlier. Even today I wonder about this beggar and ask myself who it was that I met that day. I clearly received my reward many times over for the small kindness I offered this handicapped man. I was almost helpless when help arrived seemingly out of nowhere. Not only was I able to travel home the following day without discomfort, I experienced no back problems for the next five years.

I traveled back to Hong Kong again the following year and made a token search for the mysterious Chinese beggar. He was nowhere to be found. Perhaps he had chosen a different location or maybe he never really existed. For me, this experience will always remain as one of the great mysteries and wonderments of the universe. I have often shared this story with others, either to encourage them or simply to talk about this true miracle that I received on the other side of the world.

I was so taken by this event that upon my return home I wrote an article for a Christian newspaper. I titled the article, "Could Jesus be Chinese?" It is a question that I have frequently asked but have always come to the conclusion that Jesus has many faces. It is for each of us to recognize Him even in the most unexpected places.

The Good and not so Good Experiences

India is a country that is unique in its culture and religion. As a young boy my father took me to see two adventure films entitled, *The Tiger of Eschnapur,* and *The Indian Mausoleum.* We were both fascinated by these films as they presented an India at the time of the Sultans and the Rajas. During my first visit to this picturesque country, I was fascinated by the people, the customs and the smells of spices. Little had changed from my earlier movie memories.

After visiting several cities such as Bombay, Bangalore, and Madras, I had the wonderful opportunity to spend a few days at the small city of Trivandrum, located at the most southern tip of India. The city itself is known mainly for its quality rope making, a tradition passed on from previous generations. Instead of specialized factory facilities, a small dirt road serves as the fabricating floor. Once a rope is finished it is rolled into one large ball. Because of the weight, which can be well over a hundred pounds, two people are required to lift the large ball onto the head of a third person to be carried off for storage or sale. A local guide proudly advised me that even the largest ships are moored with the ropes made by their modest process.

Trivandrum is not a glamorous city and because of its high humidity and temperature, men usually wear a wrap around the lower portion of their bodies. This adds a more casual but traditional effect. While visiting there, however, I had the good fortune to stay at the palace of the former Maharaja of the state of Kerala. The palace had been converted into a hotel and was located on the coast of the Indian Ocean. By looking through the palm trees leading down to the water, one could see the gently curved beach below. With the sun reflecting on the blue ocean, it was clearly one of the most delightful places I had ever encountered. At night from my hotel patio I would look at the stars, which seemed particularly bright. With the faint sound of instruments playing the traditional Indian rhythmic music combined with the reflection of the Indian Ocean below, I found myself transported back to the time of the thousand and one nights. I considered this city to be one of the most intriguing of the many wonderful and memorable places I was destined to enjoy. It made me realize that none of this just happened by accident or coincidence but that it was, in fact, part of my designated journey. I accepted and enjoyed being guided and led by something far greater than my simple imagination. My lifelong desire to travel and to experience exotic countries and adventures had become a reality.

Most of us have these great experiences in one form or another. Often we fail to recognize that they are a gift. When witnessing the beauty of nature or something as natural as a sunset, we come to realize God's omnipotence. We need to take time to enjoy these special moments as they are truly His gifts for us to enjoy.

Anyone who travels realizes that it can be fun and enjoyable, but occasionally can also contain unusual surprises. One of my somewhat different travel experiences occurred during a return trip to India. Together with an Indian associate, I traveled to the city of Agra, home of the famous mausoleum, the Taj Mahal. Upon our arrival at the gate leading into the park which surrounds the famous monument we noticed an unusual gathering. The crowd consisted of many peasants and more were arriving all the time. I learned from my Indian associate that the gathering was intended to be a planned riot. Within minutes, police personnel carrying long sticks arrived to control the ever increasing crowd.

We also learned that because of the potential dangers resulting from such riots, the gate to the famous grounds was temporarily locked with security personnel stationed behind it. Here we were, two men dressed in business suits in the midst of an impending riot. From my previous studies about India I recalled that these riots often turned brutal and even deadly. I also realized that India's caste system could now work to our disadvantage especially because we clearly stood out from the traditional dress of the lower class. Could we suddenly become the objects of their anger? Would the mob surge on us before the police were able to intervene? Would the police even intervene? All these questions were racing through my mind as we sought an escape.

More police and peasants arrived in the backs of trucks. The police formed a line that began to move towards the protesters who were holding signs that I was unable to read. We retreated until our backs touched the iron gate at the entrance to the park. In the meantime, my Indian associate continued to negotiate with the security guard behind the gate telling him repeatedly that we were both US citizens. The police line was now forcing the rioters in our direction. I prepared myself for the worst. How were we going to defend ourselves against so many angry demonstrators?

The police were now beginning to swing their long sticks at the demonstrators. Just as we were about to be pushed against the iron fence behind us by the approaching crowd in front of us, the officer behind the gate finally realized our predicament. He opened the gate just a few inches while shouting that anybody with a foreign passport could come in. There were only the two of us and we quickly slipped through the gate into the safety of the park. From the secure spot behind the gate we witnessed police and demonstrators now mingling in violent clashes with neither side willing to retreat. Despite the rioting outside the gate, we moved on to visit the famous Taj Mahal. The beauty of the grounds and of the monument helped us to temporarily forget about the violence and the danger we had just escaped.

Several hours later we learned that several of the demonstrators had been hospitalized for the treatment of wounds they suffered during the riot. I realized that once again I had been miraculously spared from harm. Was it just luck again that enabled us to escape this dangerous event at the very last moment? Although I will never know for certain, I nonetheless thank my guardian companion.

If luck is being at the right place at the right time, than we are unlucky if we are at the wrong place at the wrong time. We use this phrase freely when in fact it has little to do with luck at all. We could say that we should specify each moment in our lives as either lucky or unlucky or something in between. We simply have to be somewhere at any moment in time. That is part of our existence. Do we ever get up in the morning wondering whether we are going to have a lucky or an unlucky day? I think that we simply have experiences of varying degrees whether that means facing traffic when it is least heavy or facing the inconvenience of a snowstorm. Our lives are made up of constant events (many of which we take for granted). When they become non-routine, then they become notable. Some we can plan for, others simply happen either because of our doing or the doings of others. Often they occur out of nowhere and beyond anybody's doing. Natural disasters are such events.

As pointed out earlier, we are quick to judge certain positive situations or outcomes as luck. We all like to be lucky regardless of the situation. Being lucky, however, usually requires certain circumstances out of the ordinary to occur. We frequently overlook unusual circumstances that should really cause us to think. In the case of the Indian riots, it was the last minute decision by the security guard that allowed us to escape. We were no doubt lucky, however, we may want to ask, what it was that inspired the guard to change his mind at the very last moment? Perhaps it was just his compassion, but we will never know. It should prompt us to think whether an unseen force caused the guard to act at just the appropriate time. While this can surely be argued, I trust in the intervention of the invisible forces that helped me to avoid personal harm during this rather dangerous event.

Recalling a Blessed Opportunity

The ability to travel has always meant a great deal to me. It is one thing to have a desire to travel, but it is quite another to have that dream fulfilled beyond all expectations.

During my forties, I joined my first Bible study group. As a result, I developed a strong interest in the Old Testament and experienced my first introduction to Jewish culture, at least as it was described in the Bible prior to the time of Jesus. From this beginning, my interest grew and I wanted to learn more about the state of Israel in today's age. What really began to interest me was the period from 1945 to 1948, the time of the creation of the current state of Israel. Reading several books about this short period spurred my interest even farther.

The saying, "be careful what you ask for, you may get it," would prove to be true. I visited Israel for the first time around 1982. With the assistance of friendships that I had previously developed, my interests about that much debated country would now enable me to experience it first hand. I visited the places of the Old and the New Testament and stood at the tomb of King David as well as the birthplace and the place of crucifixion of our Lord, Jesus. I saw the remnants from the time of the crusaders of some nine hundred years ago and the destroyed vehicles along the road between Tel Aviv and Jerusalem, reminders of the 1945 to 1948 struggle towards statehood.

To consider myself lucky would have been an understatement. To visit Israel was truly a grace as each of my desires was being fulfilled through the most unusual circumstances. My fulfillment would reach a new height when I was invited to visit the Wailing Wall. The wall is believed to be the original temple site in the old city of Jerusalem. I was aware that this was the holiest place for Jewish people around the world and an important site for the Islamic faith as well. This made a profound impact on me as a Christian recalling that much of our religious beginning leads back to this historic and holy site. While touching the old stone wall, I felt as if I had touched Heaven itself as I realized that this site had been built by God's own instructions many centuries ago.

While enjoying the moment in near disbelief, my guide and friend motioned me to enter an adjacent building which bordered directly onto the Wailing Wall. Here I was introduced to a rabbi who apparently was in charge of this most important holy site. It seemed that he had already been briefed about my being a Christian and had just wanted to meet me for reasons that were not clear to me. After this brief introduction and what seemed to be his approval, we walked inside the building in the direction of the famous wall. After passing a security guard, we were directed to descend a ladder into what appeared to be a sort of tunnel. After reaching the bottom, we found ourselves in an underground excavation where just a few suspended light bulbs illuminated the tunnel. While looking around, I realized that we were the only visitors down there. After my eyes had adjusted to the surroundings, I became aware that we were facing the old and possibly the original temple wall. By being underground the precisely hewn large stone blocks which made up the wall were perfectly preserved. Realizing that I was now facing perhaps the original portion of King Solomon's Temple was not only overwhelming but also a moment in which I felt an indescribable emotion. Jesus taught here many times during His lifetime on earth and now I was privileged to be here. I could almost hear the voice of God as He was commanding Moses on Mount Horeb (Exodus 3:5) to remove his sandals because he was standing on holy ground. At that moment I was overcome with the powerful feeling that I too was standing on holy ground.

This would remain the most memorable of all my travel experiences. It was a miracle that I was allowed to visit this historic biblical country and to visit sites not normally open to the general public. Considering that I was not even of the Jewish faith made this a real miracle. I had the opportunity to visit Israel, including the present Palestinian city of Bethlehem, on several more occasions and I thank God for these wonderful opportunities. I doubt that anyone can attribute all of this just to luck or to my own planning.

It is simply astonishing that people I had met several years before my first visit to Israel, would be the very people who were influential enough to bring this visit about. It is always amazing how situations or circumstances, including people we meet, are often the preceding element for events that will materialize at a later time. This requires real foresight that we humans simply do not have. We can only assume and trust that it is the work of the universe that helps us achieve our dreams.

A Close Call with Disaster

We may still question whether we trust that the universe is there to assist us. We have come to realize that it is difficult to answer this question with a clear yes. It is by no means something we can just ignore. Considering some of my shared experiences, I trust that we each have our own such unexplainable situations. We must admit that these events are somewhat beyond our understanding and/or our control. I believe that the universe does help us and am certain that its influence in our daily lives occurs much more frequently than we realize. Just arriving at work or school each morning is something we take for granted. We credit ourselves for simply being punctual and arriving without incidents. The little events of a single day are so numerous that we look at most of them as routine. Unless something unusual happens that attracts our attention, we are mostly unaware of them. On those occasions when we do become aware of them, we think about how they affect us personally. Occasionally the assistance or workings of the universe in our lives may be for a dual purpose and not just intended for ourselves alone, but also for the benefit of others. I like to think that the following event was just such an occasion.

Once, when I lived in an apartment complex, I had rented a garage to store my well-preserved model 944 Porsche. The garages were connected in a single row of individual units and were located opposite the apartment buildings. One morning I decided to take my treasured car for a drive. While opening the garage door, I assured myself that

The Universe, is it Guiding Our Lives?

I could safely back the car out onto the wide driveway between the buildings. With not a person in sight, I started the car and began to back out of the garage. Once I had exited far enough to clearly look in both directions, I assured myself that it was clear of cars and people and safe to continue.

I had just started to move again when suddenly a loud male voice, which seemed to come from the passenger side right next to me, screamed "stop!" I instinctively pressed the brake pedal and the car came to a halt. As I looked into my rear view mirror, I noticed an elderly lady directly behind my car. She was short and only her head and neck could be seen above the trunk of the car. She had stretched out her left hand and was touching the back of the car as if trying to stop it from moving. With my heart pounding, I quickly stepped out of the car wondering where she had come from so unexpectedly. I also looked for the man whose shout I had heard so clearly and which had caused me to slam on the brakes.

Other than the woman who stood behind my car, there was no one else in sight. The realization that I could have killed or injured someone was a shock. The lady had moved on without looking back or even saying a word. Still in shock, I watched her as if to assure myself that she was really unharmed. I wondered about the loud male voice that seemed to have come from within the car. Yet, there was nobody else around. Wherever the voice came from, it had just prevented me from seriously hurting or possibly killing someone while at the same time it served both the lady and me as well.

The event has remained in my memory and I frequently recall how close I had come to seriously hurting someone. It still makes me shudder when I think of the incident and each time I wonder about the firm male voice that commanded me to stop. Although I may never come to know its origin, I feel a tremendous gratitude for this mysterious but timely intervention.

I had an opportunity to meet the same lady some time later and learned that her husband had passed away about a year prior to this occurrence. She also talked about him in a very loving way and so I assumed him to have been a very caring husband and father. By recalling this incident, I can't help but wonder whether it may have been his voice that commanded me to stop. If we accept the idea of a community of saints, whether here on earth or in heaven, would it be so outrageous to assume that her departed husband is still watching over her safety and well-being? It was mentioned in an earlier chapter that there is a general acceptance among most Christians that our soul returns to God, The Creator, after our passing. This is of course, provided we are deserving of this grace. The soul carries with it all the feelings it has experienced, especially those of the love it has enjoyed while here on earth. To assume otherwise would render the soul without feeling. If the primary purpose of love is to take care and protect those whom we love, then I can easily accept that this lifesaving incident was an act of love by the spirit of her departed husband.

I believe that my father, too, is with the saints and so I frequently wonder about his closeness to his family. He was a gentle man who loved his family and people in general and valued the friendships of everyone in his life. Even having spent seven years as a prisoner of war and with the illness that plagued him in his final years, he still enjoyed life and the people in it. With his humble demeanor and his acceptance of life as it was dealt him, he displayed saintly qualities to me as a father and as a human being. Perhaps it was his voice that called me to stop, but I will, of course, never know that either. I do take comfort in the thought that I can still talk to him even though he is unable to respond. Who better than our own departed family members to serve God and to help us?

Love as an Important Ingredient

It is probably not unreasonable to think that in life we all have one particular goal that we hope to accomplish but that somehow always gets delayed. It tends to decline on our priority list when we begin to think about it seriously. For me this goal was the hope of writing a book. Although I had much enthusiasm for the idea, it simply seemed an impossible project. I made several starts only to find myself discouraged by the lack of my own imagination. It wasn't until I realized that I had wanted to tell my children about my formative years during the war in Germany that I finally became inspired. By making my story into a book, I could leave this information for them and for my grandchildren as well. At the same time the book would be a tribute to my mother who all alone had to provide for and protect four young children during very difficult and frightening war years.

I eventually finished the book and it was finally published in August 2006. I have frequently asked myself whether I should credit this to the unseen forces, realizing that it was a constant inner voice that kept inspiring me to continue writing. In addition to the satisfaction I received from completing this project, I can now leave my story for future generations in order to help them better understand their heritage. During the early months of 2007, I was invited to speak about my book in my home town in Germany, where it was felt that the book had immortalized the town in some way. While giving my talk to the locals at the town's museum, Mother who was ninety-five, and other family members were in attendance. It was not until then that I fully realized the impact the book had on my old hometown community and why I had felt so inspired to write it. Being able to honor Mother in the presence of her own townspeople, including the mayor and other dignitaries, gave a real meaning to the many hours I had spent writing the book.

I have received acknowledgements from readers who felt inspired by my book to write about their own stories. It is rewarding to see how, through the inspiration I received, other people's lives are touched in different ways. I have come to recognize that although we do things for our own benefit, we can also become the conduit that often touches and inspires others. While we may not always get feedback from those we have touched, we do touch people nonetheless.

I am not so sure that we need to be cautious about expressing our gratitude to the universe. We just need to be careful not to substitute it for God who is its creator. God then is the universe as His spirit dwells everywhere within it. If we show gratitude, we acknowledge that something unusual or extraordinary has occurred. Rather than accepting things as just normal, we become cognizant of the wonderful possibility that we have friends in the universe. Some may ask, "What good are they to us if we can't communicate with them?" Perhaps by calling on them we acknowledge their existence and express our gratitude. When we consider our earthly families, our love is often best expressed when we serve the needs of our family members. Love does not always need to be verbal in order for it to be real.

Love is a powerful feeling that doesn't end when someone dies. Love stays with us and with the souls that return to the Creator to become part of His spiritual world. We believe that God is love and so why would it be otherwise with the rest of the spiritual world? It is a mutual feeling that binds us for eternity. Even those who have not felt loved have someone on the other side that cares for them and even loves them. These may well be the spirits who had not loved or felt loved while on earth themselves and now realize love's importance. If we really consider heaven to be a joyous and peaceful place as we have been taught, then love is truly the main element that brings this about.

It is difficult for us to think about love other than in human terms where we frequently harbor resentments. Often we just can not forgive one another even for petty things. Or if we forgive, we just can not quite forget. It's hard for us to let go. Envy, jealousy, and resentment often cause us to act irrationally. The spirits are no longer plagued with this negativity that affects so much of our lives, provided, of course, that they are not in what we consider hell. With their positive, loving, and caring attitude, it becomes easy to understand their feelings towards us. We often only think about what we get out of their intervention in our lives. After all it seems that other than our love and prayers there is little we can offer them in return. I think that the spirits too have a desire to be close to us and to see us happy. We should accept that, while realizing that they will always act only in our best interests.

Be Careful What You Ask For

Previously, I cautioned to be careful what you ask for, for you may get it.

Although we usually say this jokingly, we also realize that asking is what we are encouraged to do. By doing so we are in fact saying, "I am asking because I believe." It was also pointed out in our presidential debate, that just because we ask, we are not usually certain whether our request will be answered or granted.

One evening, while saying my nightly prayers, the diversity of the many denominations within our Christian faith came to mind. It was just a thought and so I did not consider it very important. Something however prompted me to ask, "God why did you allow the break in the Church to occur as a result of the Reformation, dividing it into Catholics and Protestants?" Although it was meant just as an expression of thought, an inner voice immediately responded by saying "I will not tell you, but I will help you find the answer." Being somewhat skeptical about these inner voices, I simply accepted it with a smile but with no major reaction or expectations.

Strangely enough, just several days later I visited a book store searching the shelves for books about the Reformation period. I sat for hours reading about this fascinating period. After having read all the available books about the Reformation and Martin Luther, I craved to know more. On my frequent trips to Germany, I searched the bookstores for the original books written by German authors and in the local language. I realized that German authors had a reputation for preciseness and historical facts. This would provide me with not only a true perspective but it would include the ongoing German history that unfolded into the events of the times. My obsession drove me to travel endless miles while trying to visit every possible site related to Martin Luther – where he was born, where he taught, where he had to temporarily hide out and where he died. I not only wanted to learn about this important period in history, I needed to feel the surroundings and its influence on the times. More so I wanted to understand Luther's thinking and his motivations. Luckily, with the reunification of Germany, all of these important sites were now available for all to see.

Each time I came upon a new question, the answer was always the same, "I will not tell but I will show you," and each time I was led to books, people or programs that held the answers I sought. My craving for a greater understanding of this period has not diminished. While I have gained in knowledge of the historic circumstances that brought the reformation period about, I can only form my own conclusion. While we cannot understand God's reason for allowing this enormous event in history to occur, it may have well been His way of returning Christianity to its intended purpose. As a result of my own studies, history has taken on a totally new meaning for me. As wisdom is the greatest achievement we can attain, I am truly thankful for having had this small insight into so much that affected the world's direction. Be careful what you ask for, the universe may just grant your request should cause us to think.

The Universe of the Unbeliever

Here we are still faced with the same question, is the universe guiding our lives? We need to admit, that no matter how intelligent or important we think we are, we simply can not say for certain. On the other hand rejecting the idea as nonsense would be a discredit and a grave insult to God himself. It is here where believers and non-believers really part ways. Accepting the existence of spiritual beings in the universe becomes a matter of belief and trust without any scientific evidence to verify it. The dictionary defines belief in several ways. The definition: "as the conviction that something is true," may serve our purpose best. The fact that a belief can also just be someone's opinion, still makes it a belief at least to that one person who holds that opinion.

For the unbeliever it becomes a universe without a God. Without that creative and guiding entity, the universe really has no purpose which we feel will benefit humanity. While we have gained a great deal of knowledge about our solar system through our space explorations, it serves primarily the curiosity of the scientific community. Non-believers in general don't expect more from the universe other than what science has been able to provide them. As believers we look beyond the scientific results. We look and believe in the unseen spiritual life that science is less concerned about. From the non-believer's perspective there are no expectations of anything of a personal nature to come from the universe. Although we recognize the natural order that exists within the universe, its functions or purpose, however, remain unclear to us. Without trusting in that greater unseen force behind it, one can only assume that the preciseness of the universe has come about only by coincidence and over time.

Humanity too, the unbeliever assumes, has come into existence by sheer circumstance and through thousands of years of evolution. This process was influenced by the environment in which man has lived and so much of our racial differences are the long term effects due to the various climates and conditions. Somewhere along this journey, man turned into an intelligent and thinking human being. This, by luck, has set us apart and placed us above the rest of the animal kingdom.

Evolution is not debated in the scientific community rather it is based on evidence as was discussed in an earlier chapter. There are different ways to explain evolution. One way is the theory of natural selection (Darwin). Fossil evidence, biochemical similarities, vestigial organs, and DNA all indicate that humans and apes, for example, had a common ancestor. No scientist claims that man evolved from apes. While these concepts may be easy to accept by non-believers, many believers still struggle with this issue and rather hold to creationism.

Without a creator and a meaning for our existence, we need to assume that we are merely drifters without a purpose and without hope for a better life here on earth and for eternity. Because of this assumption of a rather purposeless life, we continue to perpetuate this aimlessness by imposing it on future generations and so the purposeless cycle continues.

Although this may be their reality, non believers or atheists may not necessarily be aware or concerned with the various issues pointed out in the previous chapter. Unless, however, we understand that we have been placed here for a purpose and for a role to fulfill with the hope of an eternal spiritual life to follow, then all that is really left is the here and now. From this perspective we can understand that wealth, comfort, security, recognition, and getting the most out of life become reasonable objectives. These are by no means bad objectives as they are sought by many good believers and well known religious leaders as well. When we crave these things because we assume that there is nothing else besides the here and now, then they become the controlling factors in our lives. One only needs to recall the 1929 stock market crash and the many suicides that followed to realize the reality of these controlling factors. We tend to forget that all that we have accumulated can be gone in an instant. One needs only to recall the 2005 Katrina disaster, the wild fires in California, tornados and hurricanes as well as the power of earthquakes to realize how fragile our assets and comforts really are. If all our lives and happiness are dependent only on our assets, we live very fragile lives without a cushion to fall back on. We tend to forget that we come into the world with nothing and will leave the same way. We are in a sense only temporary users of the things we so eagerly accumulate and claim as our own.

These are harsh assessments from the viewpoint of believers and yet I am certain that many of the suicides following the 1929 stock market crash were committed not only by non-believers but by believers as well. It is interesting to note that some men, who have fought and survived the ugliness of war, are suddenly devastated even to a point of committing suicide by the loss of their possessions. We have a tendency to cling to what we have gained for ourselves even to the point where we view them as constituting our real identity.

It appears that from the standpoint of the non-believer or the atheist, the universe has nothing to offer us earthlings. This perception could very well change should we encounter visible live beings from other planets in the future. Here again, we realize the concentration on physical proof as our only means of accepting the existence of life in the universe. The above mentioned views held by non-believers would probably be my own assumptions if I did not believe in the greater power of God.

It is important to realize that whether one is a believer, a non-believer, or an atheist, the choice is personal. We are not born into one of these categories nor are the categories defined by a race that we are unable to escape from or change. It does not define us as better or worse human beings but simply declares a personal choice we have elected for ourselves. The influence of our parents and our upbringing plays an important role in what lifestyle we ultimately select for ourselves. In the end, however, we are responsible for our own lives and the choices we make.

The Universe as Viewed by the Believers

As we view our differences between the believers and those who do not believe, it becomes important to realize that the universe remains unchanged regardless. It is simply our beliefs and our openness to them that become the important variable. This distinction will be helpful when we examine the perception of the believers. We have to be careful since believers can vary greatly in their practices. This is best demonstrated by the many religious groups, each of them having different viewpoints about God and their relationship with Him. Even among our Christian faiths we have a great diversity in what we believe and how we practice our faiths. This became most evident during the recent 2008 presidential candidate debates where the discussion lead to the question of whose Christianity is real and whose is possibly considered a cult. For a nation whose origin was based on religious freedom, we are often self centered and occasionally self righteous when it comes to the religious faith of others. In spite of our petty differences, we somehow need to feel that we are in the right. In the end, we will all come before the judgment of God and He alone will decide our fate. God's judgment should make us realize that believers too have their human shortcomings and are by no means perfect.

Rather than concern ourselves with the differences between the true and the borderline believers, let us follow the viewpoint of the general believers who accept the existence of God. Most Christians believe in the Holy Trinity, namely the Father, the Son and the Holy Spirit. I view God as being all inclusive, thus any references to Him shall refer to the Holy Trinity. We need to realize, however, that aside from Christianity, no other religion embraces the concept of a Trinity. Believing in God's existence as the creator of the universe is the fundamental and most important aspect as we consider the issues at hand. He is the heart and the center of all that exists. God not only created the universe but He created it to operate with unequaled preciseness. Although what He created was perfect, it was not complete without some form of life. Scientists believe that the Earth came into being some 4.5 billion years ago- a newcomer to an already existing universe. Although God created or allowed the development of everything else, human life, with its emotions and intelligence, would come the closest to His own image. Right from the beginning man proved to be weak and easily persuaded to do wrong. We know from biblical history that God was not always pleased with this final creation, beginning with Adam and Eve and later at the time of Noah, when He allowed the great flood to occur.

At least three of the world's religions- Judaism, Christianity and Islam- share the basic belief of the existence of a one and only God. We have our individual views about which of us is more favored by Him. The fact however, that we share this belief of a one and only God is unique and an important premise. We are told in both the Old and the New Testament of numerous invisible but verbal encounters people had with God, although no one has ever claimed to have seen Him. In spite of this, we accept that God exists and believe that He is the supreme spiritual being. For us this is the first and most important premise while considering our overall spiritual objective and the role of the universe.

In an earlier chapter we touched upon the existence of angels as servants and messengers of God. It was pointed out that for the angels to interact and to serve a spirit God, they too needed to be spirit beings. For our study or analysis we may consider them a second level of spirit beings. We are taught that the angels enjoy living in the presence of God in His kingdom that we refer to as Heaven.

Because of the fallen angels, with Lucifer as their leader, a division occurred in the universe into what has come to be known as heaven and hell. It may well have been the beginning of good and evil as we

The Universe, is it Guiding Our Lives?

know it today. This division gravely affects mankind which by its free will can and must choose between good and evil, and right and wrong. Lucifer and his fallen angels were forced to depart from heaven to occupy a new place called hell. Most of the world's religions understand the terms Heaven and Hell and how they relate to good and evil. For us, these terms are symbolic. As a result of this division two opposing spiritual entities now occupied the universe.

We are generally taught that we are created in the image of God. It would be blasphemous, however, to think of ourselves as His equal. It is true that we have certain abilities, such as the ability to choose which we may consider to be a godlike quality. We do, however, fall far short of being His equal. Coming into being as humans, we are taught that we have been given a soul. As we do not fully comprehend this gift, we accept it as that unseen spirit that remains with us throughout our lifetime. Since the soul comes from the creator at the time of our birth, it is likely to return to Him at the time of our death. There it remains with Him forever. We can assume then that the soul is our spiritual connection to God and by doing so acknowledge that we too have a spirituality (even though we can't fully comprehend its significance or gift).

This then brings us to an additional category besides God and his angels and the fallen angels. These are the souls of the departed which we trust to make up a spirit world and who we have previously defined as the saints and those who are becoming saints. It seems reasonable to assume that their earthly experiences will remain with them even into the spirit world. By coming into the presence of God, they not only come to know Him more intimately, they now have the ability to see things from the human as well as the spiritual perspective. With this much broader understanding, they are no longer confined to the often unnecessary tension and frustration we experience by only being able to see things from our limited earthly perspective. I envision this as a tranquil and peaceful existence emanating from God himself.

From the Catholic perspective we view our closeness to the saints as a special connection upon which we draw rather frequently. We need only to think of our reliance on Mary, the mother of Jesus, to realize this connection. Not only is she our main intercessor, she is also the patron saint of our country. One needs only to consider the millions of Rosaries that are said around the world as a daily devotion to her. Much of our reliance on the saints stems from our hope that they have a greater influence with God than we have ourselves. I think of this as being somewhat short sighted as probably more of our prayers are directed towards Mary and the saints than to God himself. Some in the Church may not welcome this observation. Since much of this has evolved through the influence of the Church over centuries, the origin of praying to the saints may no longer be so clear, but our spiritual reliance upon them exists nonetheless.

One of our special connections to the saints, in particular to Mary, is due to her apparitions. This is probably questionable to many non-Catholics. Most Catholics have a great deal of faith in these appearances. These sightings have been claimed by many people from around the world. Even the Catholic Church is careful about acknowledging them as authentic unless actual miracles have occurred that can be associated and clearly confirmed to have resulted from these appearances. These apparitions are by far the clearest evidence of our connection to spiritual beings and support our belief in the workings of the universe.

We have also touched upon the Catholic belief in purgatory, a place that is often described to the faithful as being similar to that of a temporary Hell. We realize that being in Hell is eternal where purgatory is a temporary place of cleansing before being allowed to come into the presence of God. It was also explained that the Catholic Church has lately taken a much softer approach and now wishes to refer to purgatory as a time of transformation. This, however, should cause us to question how to view this state of transformation. Is it really a place or is it just a period of mental adjustment? Since purgatory has been defined as neither Heaven nor Hell, we have to assume it to be a separate place and so it adds to the unseen life in the universe.

Thus far we have touched upon three separate entities, namely Heaven, Hell, and purgatory. It may be that each of these so called places is able to view the others without, however, the ability "to cross over." Therein lies the misery of Hell in that it can see the glory of Heaven without any hope of ever gaining entry to it. Those who are in Heaven can witness the misery of Hell and so have an even greater awareness and appreciation as spiritual beings. Finally, assuming that purgatory in fact is a place, those souls who reside there can view both

The Universe, is it Guiding Our Lives?

Heaven and Hell and so patiently wait until they too are allowed to partake of the joys of coming into the presence of God.

As humans and sinners we will someday face these places as well. As believers we are hopeful that we too will be allowed to live in the presence of God and in the communion of the saints. Although we have hope and assurance, we struggle constantly to overcome our temptations. As a result we find ourselves vacillating between good and bad, we can feel both nudging us. Despite being aware of the deviousness of the satanic spirits, we would probably stand little chance of resisting if it were not for the positive spiritual interventions. As temptations are always lurking, this is the time when the spirits are hardest at work on our behalf. To protect us from these temptations becomes their most important duty.

This is not a point to be taken lightly when we consider the many millions of people who have lost their lives because of ruthless dictators, wars and ethnic cleansing. These evils are not just those that have occurred in the past but also those still happening in many parts of the world today. These evils are so determined that even our Church carries stains of guilt and periods of departure from its intended mission to serve God and His flock. The apologies expressed by Pope John Paul II to the Jewish leaders during his visit to Israel were not without a purpose. For us to think that we, as individuals, can continually resist such strong evil forces without a spiritual support would be gross self deception.

Our media is the greatest source of negative news. Much of this, is of course, driven by ratings. The well known TV reporter Anderson Cooper, during a 2007 presentation at the University of Buffalo, was asked why the media presented so little positive news about the events in Iraq. He explained that his broadcasting system had tried reporting good news briefly but that their ratings began to drop immediately. Naturally they continued to broadcast the more negative aspects and so regained their ratings. Anyone who watches the daily news will realize that the negative news far exceeds the positive. We should ask ourselves what that says about us on the receiving end that makes us so much more receptive to negative reporting. Constant negative reporting leads to feelings of hopelessness and despair. This is the very phenomenon that brings about discontent and unrest as well as fear in society. It occasionally allows men such as Hitler and Stalin, for example, to rise to power. The evil unleashed by these men and the many millions of lives lost because of their cruelties are all too well known.

We may claim that these atrocities have just occurred in the past and yet they continue under the guise of ethnic cleansing. Although we may not commit these heinous acts ourselves, the constant reporting of these gruesome events causes us to become hardened, and by our silence even accepting. The famous saying, "For evil to succeed requires only for good men to do nothing," should really cause us to think. Who among us has not found himself guilty of this inactivity?

Do we still doubt the evil force which seems to have such a hold on us? Denying the existence of Satan and his spiritual forces in our lives is his greatest victory. One of the great tragedies of our time is the fact that this dark force is gaining influence in our society. Our ability to even mention God in many of our public places has become forbidden. With our constant wants for more and better, we are often an easy prey. Our selfishness alone makes us easy targets. Selfishness is a force that we are not always prepared to resist. God knew this from the beginning and was not going to leave us defenseless. He requires us to have the will to resist. With our determination to resist selfishness, His spiritual beings stand ready to reinforce us. Even when we fail, we are always welcomed back to Him. God does this out of love and concern for us. Satan does not hold these values.

I have heard the thought expressed by many others that we continue to have a relationship with our dearly departed. There is a bond that is love - a feeling that we can no more explain than we can explain the existence of spiritual beings. Perhaps love is a spiritual phenomenon since it is not limited by boundaries or distances. It is like our soul, invisible but always with us. I trust that when it is our time to depart from this earth, that our loved ones will greet us and usher us into our new spiritual life. Frequently we hear of a dying person having a vision of a previously departed loved one just prior to or during their own moment of death. I believe in these occurrences and am, in fact, comforted by them. It appears that this too is one of the spirit's joyful duties - to guide us across the threshold of eternity.

Finally, for me, perhaps the most convincing evidence of a supernatural force working in my life, are my own experiences. By sharing several of my own unusual experiences in previous chapters, I have attempted to point out the unique ways in which I have benefited from this universal touching upon my life. These have been by far the most convincing situations whereby spiritual interventions have turned into physical results. I realize that not everyone will or can accept my views on this sensitive issue and that is fully understandable. Every one of us perceives circumstances differently. That is not objectionable as it simply demonstrates that we are unique in our perceptions and interpretations. If by chance, my revelations provoke the reader to consider the possibility of a supernatural force, then all the time spent writing this book will have been worthwhile.

In the preceding chapters we have touched on several instances where we as believers, feel a spiritual connection to those with whom we share the universe. Through prayer and by our petitions for intercessions by the saints, we are already acknowledging a spiritual existence. The spirits, whether we refer to them as saints or simply as spirits, are in fact, for many of us a means of communication to God. By realizing the various means in which we recognize God as our Creator and the Supreme Being and those spirits in His presence, we can begin to see the universe come alive.

Most of the practices and beliefs described in the preceding paragraphs have been in use, at least by the Catholic Church, for some two thousand years. They have evolved over time and are considered to be long held traditions of the Church and so have become part of its doctrines. By engaging and believing in these practices, we find that accepting the idea that the universe is guiding or helping our lives should come as no surprise. Many members of the Christian faith share certain Catholic beliefs. The concept of purgatory, and the devotion to the saints (especially to Mary), however, are viewed very differently by other faith groups and therefore may not be part of their religious beliefs.

From the believer's point of view, we have much evidence of spirits, places, and experiences which should inspire us to question whether the universe is alive. None of this has meaning unless we are first of all willing to believe. We know that evil has no hold over us unless we first allow it. For the positive forces in the universe to help and guide us, we need to be open, accepting, and believing.

Epilogue

Whether the universe is in some way guiding our lives will remain an open question for each of us to decide individually. We usually look to the scientific community to provide us with answers based on physical evidence. Through space explorations and the science of astronomy, we have learned a great deal about other astronomical bodies, such as new galaxies and black holes in the universe. In spite of all we have learned, much of the universe continues to remain a mystery. It is that mystery that drives us to extend our explorations farther into space. What motivates us to explore this mystery is our natural human curiosity. We simply have a desire to know what lies beyond Earth's boundaries.

As we reflect on scientific explorations, we need to acknowledge that much wisdom has been gained that will certainly benefit humanity. It is important recognize these advances so as not to diminish the value of scientific achievements that have been gained. To validate our belief that we receive guidance from the universe, the scientific community, however, offers neither proof nor support. Since science relies on physical proof of what we believers assume to be a supernatural world, we are not likely to receive even an encouraging opinion from the scientific community.

Clearly, most scientific undertakings are driven by curiosity. This is especially true as it pertains to space exploration. We only need to look at the extraordinary achievements made over the past one hundred years. By recognizing the benefits we have gained by following our curiosity, we must admit that it has proven to be a positive human trait. The desire to know may vary among individuals; it is, nonetheless, something all of us experience. Newspapers and the media are geared to feed our curiosity about the events that occur around us constantly. While we recognize this inquisitiveness within us it should not be a surprise that we find ourselves wondering whether elements in the universe are interested in engaging with us.

Believing in spiritual beings requires trust and faith and so it is natural for us to turn to religion. Here we find a great deal of disagreement among the various faiths. By universal acceptance of a spiritual God as the creator of the universe, we would have at least a strong base from which to explore a wider range of spirituality. Our human curiosity causes us to look for proof or affirmations of such spiritual connections. We look for miracles in our lives and often travel to places of apparitions in hope of being reaffirmed or having a personal spiritual experience. This then, is an acknowledgement of our belief in a supernatural world, and of a greater spiritual existence in the universe.

There is much about which we can't even begin to speculate, or ever really be able to prove. This, however, should not deter us from considering the existence of a spiritual life in the universe. Besides our belief in God as the supreme spirit in all that exists, we are quite familiar with such concepts as Heaven and Hell. For Catholics there is the additional concept of purgatory, a place usually dreaded for centuries by the believers. We have been warned about the evils of Satan and his workings in the world. The angels serve God. The saints are our intercessors and we hope to join their communion after our own passing. It is comforting to think that our own departed stand ready to assist us. We too have a spirit soul that, although invisible, is always with us. All of this was discussed in earlier chapters and need not be revisited. When we consider all these factors, it becomes easier to accept the concept that there is life in the universe.

Each of us communicates differently to God and/or the other spiritual entities mentioned. It may just be a thought, a prayer, or a personal struggle we face. Often it becomes a one-way communication or at least so it seems. A response (if and when it comes) is usually different from our expectations and so it often goes unnoticed. For me, these communications have become an important confirmation of a supernatural power influencing my life. Much of this book has been dedicated to personal experiences which I offer as evidence of my own belief.

Some will question, what has all this proved? We have no proof that we can accept with certainty. This book is simply intended to point out the possibility that there is life in the universe. Just because it is invisible does not preclude it from existing. Many of us already rely on the supernatural- the spirits, the saints, all the way to God Himself. Students of the Christian faith may even criticize my failure to mention Jesus, who is really the center of our beliefs. To me He is real as well, but I also view Him as part of the God we regard as the Supreme Being. We need to remember that not all religions, even though they may believe in the same God, view Jesus in the same manner as Christians do.

In the end it becomes a matter of faith whether we accept the notion of the universe as our guide and helper. Religion, in particular the Catholic faith which claims to be the original and true Church of God, has already provided much of the foundation for the belief in the universe by its past practices and traditions. I trust that by looking to the universe for guidance and assistance we do not offend God whose creation the universe is. The purpose of this book is not to inspire the founding of a new religion but rather to expand on what we already believe and practice. Certainly not everybody will find a belief in the universe as a source of guidance acceptable. If the simple term "universe" becomes objectionable, I suggest that it be thought of simply as God's kingdom instead.

References

(1) Book on ESCHATOLOGY, by Cardinal Joseph Ratzinger, How to Explain Purgatory to Protestants, P-13
http://www.cin.org/users/james/files/how2purg.htm   dated 8/26/2003

Biblical References: Taken from The New King James Version

Deuteronomy	6:5
Proverbs	6:32, 8:26
Peter	2:8
Matthew	10:28
Deuteronomy	18:11
John	3:16, 15:16
John	15:16
Luke	1:26-38
Luke	1:11-38
Luke	1:19
Numbers	22:23, 22:33
1 Chronicles	21:12, 21:27, 30
Isaiah	6
Romans	1:7, 8:27
Corinthians	1:2
Revelations	4:1-11

Luke	23:43
John	14:2
John	16:23
Matthew	7:18
Exodus	7:10-12
Mark	6:34-44
John	16:23
Exodus	3:5

Printed in the United States
139804LV00003BA/41/P